Other Titles in This Series

- 88 **Craig Huneke,** Tight closure and its applications, 1996
- 87 **John Erik Fornæss,** Dynamics in several complex variables, 1996
- 86 **Sorin Popa,** Classification of subfactors and their endomorphisms, 1995
- 85 **Michio Jimbo and Tetsuji Miwa,** Algebraic analysis of solvable lattice models, 1994
- 84 **Hugh L. Montgomery,** Ten lectures on the interface between analytic number theory and harmonic analysis, 1994
- 83 **Carlos E. Kenig,** Harmonic analysis techniques for second order elliptic boundary value problems, 1994
- 82 **Susan Montgomery,** Hopf algebras and their actions on rings, 1993
- 81 **Steven G. Krantz,** Geometric analysis and function spaces, 1993
- 80 **Vaughan F. R. Jones,** Subfactors and knots, 1991
- 79 **Michael Frazier, Björn Jawerth, and Guido Weiss,** Littlewood-Paley theory and the study of function spaces, 1991
- 78 **Edward Formanek,** The polynomial identities and variants of $n \times n$ matrices, 1991
- 77 **Michael Christ,** Lectures on singular integral operators, 1990
- 76 **Klaus Schmidt,** Algebraic ideas in ergodic theory, 1990
- 75 **F. Thomas Farrell and L. Edwin Jones,** Classical aspherical manifolds, 1990
- 74 **Lawrence C. Evans,** Weak convergence methods for nonlinear partial differential equations, 1990
- 73 **Walter A. Strauss,** Nonlinear wave equations, 1989
- 72 **Peter Orlik,** Introduction to arrangements, 1989
- 71 **Harry Dym,** J contractive matrix functions, reproducing kernel Hilbert spaces and interpolation, 1989
- 70 **Richard F. Gundy,** Some topics in probability and analysis, 1989
- 69 **Frank D. Grosshans, Gian-Carlo Rota, and Joel A. Stein,** Invariant theory and superalgebras, 1987
- 68 **J. William Helton, Joseph A. Ball, Charles R. Johnson, and John N. Palmer,** Operator theory, analytic functions, matrices, and electrical engineering, 1987
- 67 **Harald Upmeier,** Jordan algebras in analysis, operator theory, and quantum mechanics, 1987
- 66 **G. Andrews,** q-Series: Their development and application in analysis, number theory, combinatorics, physics and computer algebra, 1986
- 65 **Paul H. Rabinowitz,** Minimax methods in critical point theory with applications to differential equations, 1986
- 64 **Donald S. Passman,** Group rings, crossed products and Galois theory, 1986
- 63 **Walter Rudin,** New constructions of functions holomorphic in the unit ball of C^n, 1986
- 62 **Béla Bollobás,** Extremal graph theory with emphasis on probabilistic methods, 1986
- 61 **Mogens Flensted-Jensen,** Analysis on non-Riemannian symmetric spaces, 1986
- 60 **Gilles Pisier,** Factorization of linear operators and geometry of Banach spaces, 1986
- 59 **Roger Howe and Allen Moy,** Harish-Chandra homomorphisms for $\mathfrak{p}$-adic groups, 1985
- 58 **H. Blaine Lawson, Jr.,** The theory of gauge fields in four dimensions, 1985
- 57 **Jerry L. Kazdan,** Prescribing the curvature of a Riemannian manifold, 1985
- 56 **Hari Bercovici, Ciprian Foiaş, and Carl Pearcy,** Dual algebras with applications to invariant subspaces and dilation theory, 1985
- 55 **William Arveson,** Ten lectures on operator algebras, 1984
- 54 **William Fulton,** Introduction to intersection theory in algebraic geometry, 1984

(*Continued in the back of this publication*)

Introduction to Intersection Theory in Algebraic Geometry

Conference Board of the Mathematical Sciences

CBMS

Regional Conference Series in Mathematics

Number 54

Introduction to Intersection Theory in Algebraic Geometry

William Fulton

Reprinted with corrections and updates

Published for the
Conference Board of the Mathematical Sciences
by the
American Mathematical Society
Providence, Rhode Island
with support from the
National Science Foundation

Expository Lectures
CBMS Regional Conference
held at George Mason University
June 27–July 1, 1983

1991 *Mathematics Subject Classification*. Primary 14C17, 14C15, 14C40, 14M15, 14N10, 13H15.

Library of Congress Cataloging-in-Publication Data
Fulton, William 1939–
  Introduction to intersection theory in algebraic geometry.
  (Regional conference series in mathematics, ISSN 0160-7642; no. 54)
  "Expository lectures from the CBMS regional conference held at George Mason University, June 27–July 1, 1983"—T.p. verso
  Bibliography: p.
  1. Intersection theory. 2. Geometry, Algebraic. I. Conference Board of the Mathematical Sciences. II. Title. III. Series.
QA1.R33 no. 54 [QA564]   510s [512′.33]   83-25841
ISBN 0-8218-0704-8

Copying and reprinting. Individual readers of this publication, and nonprofit libraries acting for them, are permitted to make fair use of the material, such as to copy a chapter for use in teaching or research. Permission is granted to quote brief passages from this publication in reviews, provided the customary acknowledgment of the source is given.

Republication, systematic copying, or multiple reproduction of any material in this publication (including abstracts) is permitted only under license from the American Mathematical Society. Requests for such permission should be addressed to the Assistant to the Publisher, American Mathematical Society, P.O. Box 6248, Providence, Rhode Island 02940-6248. Requests can also be made by e-mail to reprint-permission@ams.org.

© Copyright 1984 by the American Mathematical Society. All rights reserved.
Third printing, with corrections, 1996.
The American Mathematical Society retains all rights
except those granted to the United States Government.
Printed in the United States of America.
∞ The paper used in this book is acid-free and falls within the guidelines
established to ensure permanence and durability.
Visit the AMS home page at URL: http://www.ams.org/
10 9 8 7 6 5 4   03 02 01 00 99

Contents

Preface ... ix

Chapter 1. Intersections of Hypersurfaces ... 1

   1.1. Early history (Bézout, Poncelet) ... 1
   1.2. Class of a curve (Plücker) ... 2
   1.3. Degree of a dual surface (Salmon) ... 2
   1.4. The problem of five conics ... 4
   1.5. A dynamic formula (Severi, Lazarsfeld) ... 5
   1.6. Algebraic multiplicity, resultants ... 6

Chapter 2. Multiplicity and Normal Cones ... 9

   2.1. Geometric multiplicity ... 9
   2.2. Hilbert polynomials ... 9
   2.3. A refinement of Bézout's theorem ... 10
   2.4. Samuel's intersection multiplicity ... 11
   2.5. Normal cones ... 12
   2.6. Deformation to the normal cone ... 15
   2.7. Intersection products: a preview ... 16

Chapter 3. Divisors and Rational Equivalence ... 19

   3.1. Homology and cohomology ... 19
   3.2. Divisors ... 21
   3.3. Rational equivalence ... 22
   3.4. Intersecting with divisors ... 24
   3.5. Applications ... 26

Chapter 4. Chern Classes and Segre Classes ... 29

   4.1. Chern classes of vector bundles ... 29
   4.2. Segre classes of cones and subvarieties ... 31
   4.3. Intersection forumulas ... 33

Chapter 5. Gysin Maps and Intersection Rings ... 37

   5.1. Gysin homomorphisms ... 37
   5.2. The intersection ring of a nonsingular variety ... 39
   5.3. Grassmannians and flag varieties ... 41
   5.4. Enumerating tangents ... 43

Chapter 6. Degeneracy Loci ... 47

6.1.	A degeneracy class	47
6.2.	Schur polynomials	49
6.3.	The determinantal formula	50
6.4.	Symmetric and skew-symmetric loci	51

Chapter 7. Refinements — 53

7.1.	Dynamic intersections	53
7.2.	Rationality of solutions	54
7.3.	Residual intersections	55
7.4.	Multiple point formulas	56

Chapter 8. Positivity — 59

8.1.	Positivity of intersection products	59
8.2.	Positive polynomials and degeneracy loci	60
8.3.	Intersection multiplicities	62

Chapter 9. Riemann-Roch — 63

9.1.	The Grothendieck-Riemann-Roch theorem	63
9.2.	The singular case	66

Chapter 10. Miscellany — 69

10.1.	Topology	69
10.2.	Local complete intersection morphisms	70
10.3.	Contravariant and bivariant theories	71
10.4.	Serre's intersection multiplicity	74

References — 75

Notes (1983–1995) — 77

Preface

These lectures are designed to provide a survey of modern intersection theory in algebraic geometry. This theory is the result of many mathematicians' work over many decades; the form espoused here was developed with R. MacPherson.

In the first two chapters a few epsisodes are selected from the long history of intersection theory which illustrate some of the ideas which will be of most concern to us here. The basic construction of intersection products and Chern classes is described in the following two chapters. The remaining chapters contain a sampling of applications and refinements, including theorems of Verdier, Lazarsfeld, Kempf, Laksov, Gillet, and others.

No attempt is made here to state theorems in their natural generality, to provide complete proofs, or to cite the literature carefully. We have tried to indicate the essential points of many of the arguments. Details may be found in [**16**].

I would like to thank R. Ephraim for organizing the conference, and C. Ferreira and the AMS staff for expert help with preparation of the manuscript.

Preface to the 1996 printing

In this revision, we have taken the opportunity to correct some errors and misprints. In addition, a section of notes has been added, to point out some of the work that has been done since the first edition was written that is closely related to ideas discussed in the text. Superscripts in the text refer to these notes. As in the text, no attempt is made to survey the large and growing literature in intersection theory.

I am grateful to Jeff Adler for preparing and improving the manuscript and diagrams.

William Fulton
Chicago, IL
December, 1995

CHAPTER 1

Intersections of Hypersurfaces

1.1. Early history (Bézout, Poncelet)

A most basic question in intersection theory is to describe the intersection of several algebraic hypersurfaces in n-space, i.e., the common solutions of several polynomials in n variables. The ancients certainly knew about the possible intersections of lines and conics in the plane, and they also knew that rational solutions of two quadric equations in three variables behaved like solutions of one cubic equation in two variables [61].

We do not know who first observed that two plane curves of degrees p and q should intersect in pq points. By 1680 Newton [48] had developed an elimination theory for two such equations. This produced a *resultant*, which was a polynomial in one variable of degree pq whose solutions gave an abscissa of the intersection points of the two curves. The corresponding construction and assertion for n equations in n variables were made in 1764 by Bézout [5, 6]. Bézout's treatment was entirely algebraic, although he briefly interpreted his result for $n = 2$ and $n = 3$: the number of intersections of two plane curves (or three surfaces in space) is *at most* the products of their degrees.

By referring to the resultants, which are polynomials in one variable, one can also discuss the possibilities of *nonreal* solutions, *asymptotic* solutions, and *multiple* solutions. As geometry developed, the first two of these situations were subsumed by considering intersections of hypersurfaces $H_1, \ldots, H_n$ in complex projective space $\mathbb{P}^n_{\mathbb{C}}$. Now we assign an *intersection multiplicity*

$$i(P) = i(P, H_1 \bullet \ldots \bullet H_n)$$

to a point P of the intersection $\bigcap H_i$; if the H_i do not meet transversally at P, this multiplicity will be greater than one.

Although there was little early discussion of this multiplicity, the governing *principle of continuity* was well understood, at least since Poncelet [51]. If the H_i vary in families $H_i(t)$, with $H_i(0) = H_i$, and $P_1(t), \ldots, P_r(t)$ are the points of the general intersection $\bigcap H_i(t)$ which approach P as $t \to 0$, then

$$i(P, H_1 \bullet \ldots \bullet H_n) = \sum_{j=1}^{r} i\big(P_j(t), H_1(t) \bullet \ldots \bullet H_n(t)\big).$$

Varying the H_i so that the $H_i(t)$ meet transversally, this determines the multiplicity $i(P, H_1 \bullet \ldots \bullet H_n)$.

In all the above discussion, it is assumed that the intersection of the hypersurfaces is a finite set, or at least that P is an isolated point of $\bigcap H_i$.

1.2. Class of a curve (Plücker)

An important early application of Bézout's theorem was for the calculation of the *class* of a plane curve C, i.e., the number of tangents to C through a given general point Q:

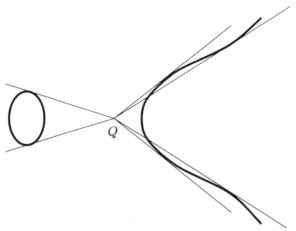

Equivalently, the class of C is the degree of the dual curve $C^\vee$. If $F(x, y, z)$ is the homogeneous polynomial defining C and $Q = (a:b:c)$, then the *polar curve* C_Q is defined by
$$F_Q(x, y, z) = aF_x + bF_y + cF_z,$$
where $F_x = \partial F(x, y, z)/\partial X$, F_y, and F_z are partial derivatives. This is defined so that a nonsingular point P of C is on C_Q exactly when the tangent line to C at P (defined by $XF_x(P) + YF_y(P) + ZF_z(P) = 0$) passes through Q. One checks that C meets C_Q transversally at P if P is not a flex on C, so
$$\mathrm{class}(C) = \#(C \cap C_Q) = \deg C \deg C_Q = n(n-1),$$
if n is the degree of C, and C is nonsingular.

If C has singular points, however, they are always on $C \cap C_Q$, so they must contribute. For example, if P is an ordinary node (resp. cusp) and Q is general, then
$$i(P, C \bullet C_Q) = 2 \quad (\text{resp. } i(P, C \bullet C_Q) = 3).$$
This gives the first Plücker formula [50]
$$n(n-1) = \mathrm{class}(C) + 2\delta + 3\kappa,$$
if C has degree n, δ ordinary nodes, κ ordinary cusps, and no other singularities.

1.3. Degree of a dual surface (Salmon)

In 1847 Salmon [53] made a similar study of surfaces. If $S \subset \mathbb{P}^3$ is a surface, the degree of the dual (or "reciprocal") surface $S^\vee$ is the number of points $P \in S$ such that the tangent plane to S at P contains a given general line ℓ. (This number is one of the projective characters of S, now called the *second class* of S.)

For a point $Q \in \mathbb{P}^3$, let S_Q be the *polar surface* of S with respect to Q: if $F(x, y, z, w)$ defines S and $Q = (a:b:c:d)$, then $aF_x + bF_y + cF_z + dF_w$ defines S_Q. Taking two points Q_1, Q_2 on ℓ, one sees as before that a nonsingular point P of S

is on $S_{Q_1} \cap S_{Q_2}$ if and only if the tangent plane to S at P contains ℓ. Thus for S nonsingular of degree n, and Q_1, Q_2 general,

$$\deg(S^\vee) = \#(S \cap S_{Q_1} \cap S_{Q_2}) = n(n-1)^2.$$

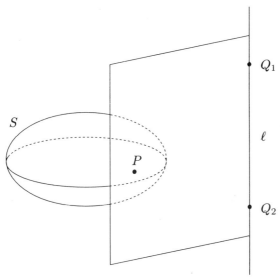

As before, all singular points of S are contained in $S \cap S_{Q_1} \cap S_{Q_2}$. If P is an isolated singular point of S, its contribution to the total $n(n-1)^2$ is the intersection multiplicity $i(P, S \bullet S_{Q_1} \bullet S_{Q_2})$. For example, the contribution of an ordinary double point is two, so $\deg(S^\vee) = n(n-1)^2 - 2\delta$ if S has δ ordinary double points.

If S is singular along a curve C, however, a new phenomenon occurs, a problem of *excess intersection*: how to compute the contribution of C to the total intersection $n(n-1)^2$, so that $n(n-1)^2$ diminished by this contribution, and by contributions of other singular points, yields $\deg(S^\vee)$. Salmon initiates a study of the contribution of a curve C to the intersection of three surfaces in space when C is a component of their intersection. For example, if C is a line, he gives its contribution as $m+n+p-2$, where m, n, p are the degrees of the surfaces. Salmon justifies this by saying that the answer must be independent of the choice of surfaces of given degrees, and then he calculates it directly in the degenerate case when the first is the union of a plane containing C and a general surface of degree $m-1$. This surface meets the other two surfaces in $(m-1)np$ points, $m-1$ of which are on the line C. The plane meets the other two surfaces in curves of degrees $n-1$ and $p-1$ in addition to C; these curves meet in $(n-1)(p-1)$ points. The total number of points of intersection outside C is therefore

$$(m-1)np - (m-1) + (n-1)(p-1) = mnp - (m+n+p-2),$$

as asserted. In case C is a double line on the first surface, he calculates its contribution as $m + 2n + 2p - 4$ by working out the case where this surface is the union of two surfaces containing C.

If C is a double line on a surface S of degree n, this analysis predicts $5n - 8$ as the contribution of C to the intersection of S with S_{Q_1} and S_{Q_2}. However, as Salmon points out, there are special points on C, called *pinch points* (or "cuspidal" points), where the two tangents planes to S coincide.

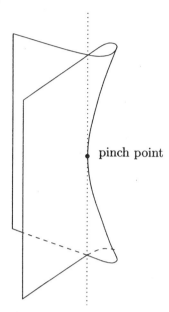
pinch point

If C is the line $x = y = 0$, and S is the surface $Ux^2 + Vxy + Wy^2 = 0$, then these pinch points are the intersections of C with the surface $V^2 = 4UW$, so there are $2n - 4$ pinch points on S. Thus C, together with its pinch points, diminishes the degree of $S^\vee$ by $(5n - 8) + (2n - 4) = 7n - 12$. For example, a cubic with a double line (e.g. $y^2 = zx^2 + x^3$) has a dual surface of degree three.

Salmon also considers more general curves. If C is a complete intersection of surfaces of degrees a and b, and C is a component of intersection of three surfaces of degrees m, n, and p, then he finds that the contribution of C to the total number of mnp is $ab(m + n + p - (a + b))$. Concluding this remarkable paper, he deduces that if such C is an r-fold curve on a surface S, then it diminishes the degree of the dual by

$$ab\big[(r-1)(3r+1)n - r^2(r-1)(a+b) - 2r(r-1)\big].$$

1.4. The problem of five conics

Problems of excess intersection arise frequently in enumerative problems. The famous problem of the number of plane conics tangent to five given conics in general position is a typical example of this. A plane conic is defined by a quadratic polynomial $ax^2 + by^2 + cxy + dx + ey + f$, unique up to multiplication by a nonzero scalar, so the space of conics can be identified with $\mathbb{P}^5$. One checks that the condition to be tangent to a fixed nonsingular conic is described by a hypersurface of degree six in $\mathbb{P}^5$. The desired conics are then represented by the points in the intersection of five such hypersurfaces $H_1 \cap \cdots \cap H_5$. There are not $6^5 = 7776$ such conics, however, as originally thought by Steiner and others. Indeed, the Veronese surface $V \cong \mathbb{P}^2$ of conics which are double lines is contained in $\bigcap H_i$, and one can show (cf. §4 below) that the contribution of V to the intersection is actually 4512, which leaves 3264, the actual number of (nonsingular) conics tangent to five given conics in general position.

Note that the conics tangent to a fixed line form a quadric hypersurface in $\mathbb{P}^6$. Given five general lines, the Veronese contributes 31 to the predicted intersection number 2^5 for the five quadrics. Since everyone knew that there is only one nonsingular conic tangent to five general lines (by duality, for example), it is curious that these false answers were proposed when the lines are replaced by curves of higher degree.

In spite of the clear exposition of the importance of excess intersections in enumerative geometry by Salmon and Cayley, such considerations played little role in the great development of enumerative geometry at the hands of Chasles, de Jonquières, Schubert, Halphen, Zeuthen, and others. For one thing, they avoided writing equations for varieties and, especially, for parameter spaces. In general, however, their work can be interpreted as calculating intersections on appropriate spaces so that the intersections become proper. Often these spaces are blow-ups of the naïve spaces, which amounts to adding structure to degenerate figures. For example, a classical approach to the space of conics amounts to working on the space of *complete* conics, which is the blow-up $\widetilde{\mathbb{P}}^5$ of $\mathbb{P}^5$ along the Veronese; in this model a point in the exceptional divisor corresponds to a double line together with a pair of points on the line. The proper transforms of the hypersurfaces H_i then meet properly on $\widetilde{\mathbb{P}}^5$ outside the exceptional divisor, and once one knows an appropriate "intersection ring" for $\widetilde{\mathbb{P}}^5$ one may calculate their intersection.

The same approach works for quadrics of arbitrary dimension. The beautiful study of complete quadrics was initiated by Schubert, who found many enumerative formulas. The rigorous construction of these parameter spaces and their intersection rings has been carried out by Semple and Tyrell, with modern re-examination by Vainsencher, Laksov, and Lazarsfeld. Realizing the spaces as orbit spaces of suitable group actions, by Demazure and by De Concini and Procesi, has led to a clearer understanding of their structure.[1]

1.5. A dynamic formula (Severi, Lazarsfeld)

In general, if $H_1, \ldots, H_n$ are arbitrary hypersurfaces in $\mathbb{P}^n$, with $d_i = \deg(H_i)$, Severi [58] proposed to assign numbers $i(Z)$ to certain distinguished subvarieties Z of the intersection locus $H_1 \cap \cdots \cap H_n$, so that

$$\sum i(Z) = d_1 \cdot \ldots \cdot d_n.$$

Each irreducible component of $\bigcap H_i$ should be distinguished, and each isolated point should be assigned its intersection multiplicity. In general, as in Salmon's examples, there may also be embedded distinguished varieties. Severi's dynamic procedure, corrected and completed by Lazarsfeld [40], can be summarized as follows. If F_i is a homogeneous equation for H_i, consider deformations $H_i(t)$ of H_i defined by homogeneous polynomials $F_i + tG_i + t^2 G'_i + \cdots$. For a given subvariety Z of $\bigcap H_i$, let $j(Z)$ be the number of points of $\bigcap H_i(t)$ which approach Z as $t \to 0$, for a *generic* deformation; in fact, $j(Z)$ of the points will have limits in Z for any deformation whose first order parts $(G_1, \ldots, G_n)$ belong to a certain open set U_Z of the space of n-tuples of polynomials of degrees $d_1, \ldots, d_n$. For any point P set $i(P) = j(P)$. Only finitely many points will have $i(P) \neq 0$. For an irreducible curve C, set

$$i(C) = j(C) - \sum_{P \in C} i(P),$$

so $i(C)$ is the number of points that generically approach C, but not any particular point on C. Inductively,
$$i(Z) = j(Z) - \sum i(V),$$
the sum over all proper irreducible subvarieties V of Z. Then $\sum i(Z) = d_1 \cdot \ldots \cdot d_n$, which achieves the desired decomposition.

We will later see a *static* construction of this decomposition, which is also valid in contexts where such deformations are unavailable. It should be emphasized, however, that in spite of the existence of a rigorous general theory, and some explicit formulas, the actual computation of the contributions $i(Z)$ remains a difficult problem.

For plane curves, following Segre [55], Lazarsfeld gives the following answer. If $H_i = D_i + E$, where D_1 and D_2 meet properly, and P is a point in E, let G_i be generic as above, let A_i be equations for D_i, and let F be the curve defined by $A_1 G_2 - A_2 G_1$. Then
$$i(P) = i(P, E \bullet F) + i(P, D_1 \bullet D_2).$$

For example, if $H_1 = 2L_1 + L_2$, $H_2 = L_1 + 2L_2$, with L_1, L_2 lines meeting at a point P, then the Segre-Lazarsfeld formula shows that
$$i(P) = i(L_1) = i(L_2) = 3.$$

1.6. Algebraic multiplicity, resultants

For an isolated point P in the intersection of hypersurfaces $H_1, \ldots, H_n$ in $\mathbb{P}^n$, a modern *static* definition of the intersection multiplicity is
$$i(P, H_1 \bullet \ldots \bullet H_n) = \dim_{\mathbb{C}} \mathcal{O}_P/(f_1, \ldots, f_n),$$
where $\mathcal{O}_P$ is the local ring of $\mathbb{P}^n$ at P, and f_i is a local equation for H_i in $\mathcal{O}_P$. If P is the origin in $\mathbb{C}^n \subset \mathbb{P}^n$, $\mathcal{O}_P$ is the localization of $\mathbb{C}[X_1, \ldots, X_n]$ at the maximal ideal $(X_1, \ldots, X_n)$. Or one may replace $\mathcal{O}_P$ by its completion $\mathbb{C}[[X_1, \ldots, X_n]]$, or by the ring $\mathbb{C}\{X_1, \ldots, X_n\}$ of convergent power series. This algebraic construction of intersection multiplicity dates from Macaulay [41].

Let us verify the agreement of this definition with that obtained from elimination theory, at least for plane curves. Suppose the curves are defined by polynomials $f(x, y)$ and $g(x, y)$, and the two curves do not meet at infinity on the y-axis. Thus we may assume
$$f(x, y) = a_0(x) y^n + a_1(x) y^{n-1} + \cdots + a_n(x)$$
with $a_0(0) \neq 0$. Let $A = \mathbb{C}[x]_{(x)}$ be the local ring of the x-axis at the origin. Then $A[y]/(f)$ is an A-algebra which is a free A-module of rank n, and one may construct the *resultant* $r = R(f, g)$ in A by
$$r = \det \left(A[y]/(f) \xrightarrow{\cdot g} A[y]/(f) \right).$$

(It is a formal exercise, left to the reader, to show that this agrees with the usual definition, as in [60].)

We must show that the order of vanishing of r at $x = 0$ is equal to the sum of the intersection numbers of the two curves at all points P on the y-axis:

1.6. ALGEBRAIC MULTIPLICITY, RESULTANTS

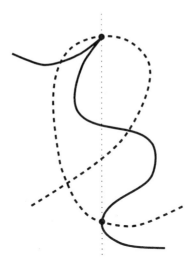

Now $A[y]/(f,g)$ is finite dimensional over $\mathbb{C}$, so it is a direct sum of its localizations $\mathcal{O}_P/(f,g)$, where P varies over the points on the y-axis on both curves. Therefore

$$\sum_P i(P) = \dim_{\mathbb{C}} A[y]/(f,g).$$

Since the order of vanishing of r at $x = 0$ is $\dim_{\mathbb{C}} A/(r)$, the equation to be proved is

$$\dim_{\mathbb{C}} A[y]/(f,g) = \dim_{\mathbb{C}} A/(r).$$

This is a special case of an important algebraic fact:

LEMMA. *Let A be a one-dimensional Noetherian local domain, M a finitely generated free A-module and $\phi: M \longrightarrow M$ an A-homomorphism. Then*

$$\mathrm{length}_A(M/\phi(M)) = \mathrm{length}_A(A/(\det(\phi))).$$

The *length* of an A-module N is d if there is a chain of submodules $N = N_0 \supset N_1 \supset \cdots \supset N_d = 0$, where N_i/N_{i+1} is isomorphic to the residue field of A. In case A contains a subfield K which maps isomorphically to its residue field, then $\mathrm{length}_A N = \dim_K N$.

When A is a discrete valuation ring, the lemma is an exercise in elementary divisors. For the general case see [**16**, A2.6].

CHAPTER 2

Multiplicity and Normal Cones

2.1. Geometric multiplicity

A *subvariety* X of $\mathbb{C}^N$ is defined by a prime ideal $I(X)$ in $\mathbb{C}[X_1, \ldots, X_N]$. The *coordinate ring* $\Gamma(X)$ is the residue ring

$$\Gamma(X) = \mathbb{C}[X_1, \ldots, X_N]/I(X).$$

A (closed) *subscheme* Z of X is determined by an ideal $I = I(Z)$ of $\Gamma(X)$, which is a subvariety if $I(Z)$ is prime. In this case the *local ring of X at Z* is the localization of $\Gamma(X)$ at $I(Z)$, and is denoted $\mathcal{O}_{Z,X}$.

If Z is a subscheme of X, the *irreducible components* of Z are the subvarieties of X corresponding to the minimal prime ideals of $\Gamma(X)$ that contain $I(Z)$. If V is such a component, the *geometric multiplicity of V in Z* is defined to be the length of the Artinian ring

$$\mathcal{O}_{V,Z} = \mathcal{O}_{V,X}/I(Z)\mathcal{O}_{V,X}.$$

The *cycle of Z*, denoted $[Z]$, is defined to be the formal sum

$$[Z] = \sum_{i=1}^{r} m_i[V_i],$$

where $V_1, \ldots, V_r$ are the irreducible components of Z, and m_i is the geometric multiplicity of V_i in Z. For example, if $X = \mathbb{C}^n$ and Z is the scheme-theoretic intersection of n hypersurfaces that meet properly, then

$$[Z] = \sum i(P)[P],$$

the sum over the points P in Z, with $i(P)$ the intersection number described in §1.6.

For an arbitrary variety X, subschemes Z are defined by ideal sheaves $\mathcal{I} = \mathcal{I}(Z)$. On any affine open $U \subset X$ that meets Z, $\mathcal{I}$ is given by an ideal in the coordinate ring of U, which is prime if Z is a subvariety. The local ring of X along V, and the geometric multiplicity of a component V of Z can be defined using any such U.

2.2. Hilbert polynomials

A subscheme Z of $\mathbb{P}^N$ is defined by a homogeneous ideal $I = I(Z)$ in $\mathbb{C}[X_0, \ldots, X_N]$. If $\mathbb{C}[X_0, \ldots, X_N]_t$ denotes the homogeneous polynomials of degree t, such an ideal I is the direct sum of its intersections I_t with $\mathbb{C}[X_0, \ldots, X_N]_t$. Two homogeneous ideals define the same subscheme when their homogeneous pieces are the same for all but finitely many t. The *Hilbert polynomial* of Z is the polynomial $P_Z(t)$ such that

$$P_Z(t) = \dim_{\mathbb{C}}(\mathbb{C}[X_0, \ldots, X_N]_t/I_t)$$

for all sufficiently large t. Indeed, one shows (cf. [**30**, §1.7]; or [**57**]) that the right side is a polynomial of degree equal to the dimension of Z, for $t \gg 0$. If $n = \dim(Z)$, one may define the *degree* of Z, $\deg(Z)$, to be the coefficient of $t^n/n!$ in $P_Z(t)$, i.e.

(i) $$P_Z(t) = \deg(Z)t^n/n! + \text{lower terms}.$$

It also follows that if $[Z] = \sum m_i[V_i]$ is the cycle of Z, then

(ii) $$\deg(Z) = \sum_{\dim(V_i)=n} m_i \deg(V_i).$$

If V is a subvariety of $\mathbb{P}^N$, and H is a hypersurface of degree m in $\mathbb{P}^N$ not containing V, then

(iii) $$\deg(V \cap H) = m \deg(V).$$

It will later become clear that this definition of $\deg(V)$ agrees with the geometric notion of counting intersections of V with complementary linear spaces. In fact, we shall have no need for Hilbert polynomials, although they have played an important role in the modern algebraic development of multiplicity.

2.3. A refinement of Bézout's theorem

The elementary facts about degree in the preceding section, together with an important join construction, allow a simple proof of the following proposition. A stronger result will appear later when more intersection theory is available.

PROPOSITION. *Let $V_1, \ldots, V_s$ be subvarieties of $\mathbb{P}^N$, and let $Z_1, \ldots, Z_r$ be the irreducible components of $V_1 \cap \cdots \cap V_s$. Then*

$$\sum_{i=1}^{r} \deg(Z_i) \leq \prod_{j=1}^{s} \deg(V_j).$$

PROOF. By a simple induction, one may assume $s = 2$. Construct the *ruled join* $J = J(V_1, V_2)$ in $\mathbb{P}^{2N+1}$ as follows. Let $X_0, \ldots, X_N, Y_0, \ldots, Y_N$ be homogeneous coordinates on $\mathbb{P}^{2N+1}$. Let $\mathbb{P}_1^N$ (resp. $\mathbb{P}_2^N$) be the linear subspace of $\mathbb{P}^{2N+1}$ defined by the vanishing of all Y_i (resp. all X_i). Identifying $\mathbb{P}_i^N$ with $\mathbb{P}^N$, one has $V_i \subset \mathbb{P}_i^N$. Let J be the union of all lines from points of V_1 to points of V_2. Algebraically, the homogeneous coordinate ring of J is simply the tensor product of the homogeneous coordinate rings of V_1 and V_2. One verifies that

(i) $$\deg(J) = \deg(V_1) \deg(V_2).$$

Let L be the linear subspace of $\mathbb{P}^{2N+1}$ defined by $X_i = Y_i$, $0 \leq i \leq N$. Then $L = \mathbb{P}^N$ and

(ii) $$L \cap J = V_1 \cap V_2.$$

Thus we are reduced to the case where one of the varieties being intersected is a linear subspace.

Since a linear subspace is an intersection of hyperplanes, one is further reduced inductively to the case where one of the varieties, say V_1, is a hyperplane. In this case, either $V_1 \supset V_2$ and the proposition holds with equality, or $[V_1 \cap V_2] = \sum_{i=1}^{r} m_i [Z_i]$, where the Z_i are the irreducible components of $V_1 \cap V_2$, and by (ii) and (iii) of §2.2 (for any hypersurface V_1 not containing V_2),

$$\sum m_i \deg(Z_i) = \deg(V_1) \deg(V_2). \quad \square$$

2.4. Samuel's intersection multiplicity

Suppose $H_1, \ldots, H_n$ are hypersurfaces in an n-dimensional variety V, and P is an isolated point of $\bigcap H_i$. Let $A = \mathcal{O}_{P,V}$ be the local ring of V along P, and assume each H_i is defined by one element f_i in A. Let $I = (f_1, \ldots, f_n)$. Then A/I is finite dimensional over $\mathbb{C}$, and if P is a nonsingular point of V, one may use $\dim_{\mathbb{C}} A/I$ to give a workable definition of the intersection multiplicity $i(P, H_1 \bullet \ldots \bullet H_n)$ as in §1. The following is a standard example of the failure of this definition in general.

EXAMPLE. Let V be the image of the mapping $\phi \colon \mathbb{C}^2 \longrightarrow \mathbb{C}^4$ defined by $\phi(s,t) = (s^4, s^3t, st^3, t^4)$, let P be the origin, and let H_1 and H_2 be the hypersurfaces of V defined by the coordinates x_1 and x_4 respectively. By varying H_1 and H_2, the principle of continuity requires that the intersection multiplicity is 4. However, one calculates that the ideal of V is generated by $x_1x_4 - x_2x_3$, $x_1^2x_3 - x_2^3$, $x_2x_4^2 - x_3^3$, and $x_2^2x_4 - x_3^2x_1$, from which it follows that $\dim_{\mathbb{C}} A/(x_1, x_4) = 5$.

Samuel [54] defines the multiplicity $i(P) = i(H_1 \bullet \ldots \bullet H_n)$ to be the coefficient of $t^n/n!$ in the Hilbert-Samuel polynomial

(i) $$P(t) = \dim_{\mathbb{C}}(A/I^t) = i(P)t^n/n! + \text{lower terms}$$

for $t \gg 0$. To see that $\dim(A/I^t)$ is a polynomial of degree n in t, for $t \gg 0$, one may proceed as follows. Let $\Lambda = A/I$ and consider the surjection of graded rings

(ii) $$\Lambda[X_1, \ldots, X_n] \longrightarrow \bigoplus_{t=0}^{\infty} I^t/I^{t+1}$$

which maps X_i to the image of f_i in I/I^2. The kernel of this homomorphism is a homogeneous ideal which defines a subscheme $\mathbb{P}(C)$ of projective $(n-1)$-space $\mathbb{P}_\Lambda^{n-1}$ over Λ. (Those who feel uncomfortable with projective space over a ring such as Λ may realize $\mathbb{P}(C)$ in $\mathbb{P}^{n-1} \times V$, since Λ is a residue ring of A.) This scheme $\mathbb{P}(C)$ is the *projective normal cone* to $\bigcap H_i$ in V. We shall discuss normal cones in succeeding sections. Here we shall use the fact that $\mathbb{P}(C)$ has pure dimension $n-1$, so its Hilbert polynomial has the form

(iii) $$P_{\mathbb{P}(C)}(t) = \dim_{\mathbb{C}} I^t/I^{t+1} = i(P)t^{n-1}/(n-1)! + \cdots$$

for $t \gg 0$. A simple calculation shows that this definition of $i(P)$ is the same as that in (i). However, since $\mathbb{P}(C) \subset \mathbb{P}_\Lambda^{n-1}$, the only component of $\mathbb{P}(C)$ is the underlying variety $\mathbb{P}_{\mathbb{C}}^{n-1}$ of $\mathbb{P}_\Lambda^{n-1}$ and, therefore,

(iv) $$[\mathbb{P}(C)] = i(P)[\mathbb{P}_{\mathbb{C}}^{n-1}]$$

defines the multiplicity $i(P)$ without reference to Hilbert functions. In addition, since $\mathbb{P}(C) \subset \mathbb{P}_\Lambda^{n-1}$, and

$$[\mathbb{P}_\Lambda^{n-1}] = \dim_{\mathbb{C}}(\Lambda)[\mathbb{P}_{\mathbb{C}}^{n-1}],$$

it follows that

(v) $$i(P) \leq \dim_{\mathbb{C}}(\Lambda) = \dim_{\mathbb{C}} A/(f_1, \ldots, f_d).$$

We see also that equality holds in (v) if the morphism (ii) is an isomorphism. This is related to the important notion of a regular sequence.

DEFINITION. A sequence of elements $f_1, \ldots, f_d$ in the maximal ideal of a local ring A is a *regular sequence* if f_1 is a non-zero-divisor in A, and if, for $i = 2, \ldots d$, the image of f_i in $A/(f_1,\ldots,f_{i-1})$ is a non-zero-divisor. (This is equivalent to asserting that the Koszul complex

$$0 \longrightarrow \bigwedge\nolimits^d(A^d) \longrightarrow \bigwedge\nolimits^{d-1}(A^d) \longrightarrow \cdots \longrightarrow A^d \longrightarrow A$$

defined by $f_1,\ldots,f_d$ is exact, giving a resolution of A/I. In fact, the multiplicity $i(P)$ may also be defined to be the alternating sum of the dimensions of the homology groups of this complex, cf. [57].)

The *dimension* of a local ring A is the length n of a maximal chain of prime ideals $P_0 \subsetneq P_1 \subsetneq \cdots \subsetneq P_n \subsetneq A$. If A is the local ring of a variety V along a subvariety W, the dimension of A is the codimension of W in V. The ring A is *Cohen-Macaulay* if its maximal ideal contains a regular sequence of $\dim(A)$ elements. For example, if P is a nonsingular point of V, then $\mathcal{O}_{P,V}$ is Cohen-Macaulay.

The following lemma contains the main facts from commutative algebra that we will need. For proofs, see [38] or [57].

LEMMA. *(a) If A is Cohen-Macaulay, a sequence $f_1,\ldots f_d$ of elements in the maximal ideal of A is a regular sequence if and only if*

$$\dim(A/(f_1,\ldots,f_d)) = \dim(A) - d.$$

(b) Let $f_1,\ldots,f_d$ be a regular sequence in a local ring A, and let $I = (f_1,\ldots f_d)$. Then the canonical homomorphism

$$A/I[X_1,\ldots,X_d] \longrightarrow \bigoplus_{t=0}^{\infty} I^t/I^{t+1},$$

which takes X_i to the image of f_i in I/I^2, is an isomorphism. Moreover, the kernel of the canonical surjection

$$A[X_1,\ldots,X_d] \longrightarrow \bigoplus_{t=0}^{\infty} I^t$$

is generated by the elements $f_i X_j - f_j X_i$, $1 \le i < j \le d$.

For example, with notation as at the beginning of this section, if $\mathcal{O}_{P,V}$ is Cohen-Macaulay, it follows that

$$i(P, H_1 \bullet \ldots \bullet H_n) = \dim_{\mathbb{C}} \mathcal{O}_{P,V}/(f_1,\ldots,f_n),$$

i.e., Samuel's sophisticated multiplicity agrees with the naïve multiplicity of §1.

2.5. Normal cones

If W is a subscheme of an affine variety V, defined by an ideal I in the coordinate ring A of V, the *normal cone* $C = C_W V$ to W in V is defined to be

$$C = \mathrm{Spec}\left(\bigoplus I^t/I^{t+1}\right).$$

The isomorphism of the coordinate ring of W with $A/I = I^0/I^1$ determines a morphism $p_C \colon C \longrightarrow W$, called the *projection*, and a closed embedding $s_C \colon W \longrightarrow C$, called the *zero section*, with $p_C \circ s_C = \mathrm{id}_W$. If $f_1,\ldots,f_d$ generate I, the canonical surjection of $A/I[X_1, \ldots, X_d]$ onto $\bigoplus I^t/I^{t+1}$ determines a closed embedding of C in $W \times \mathbb{C}^d$:

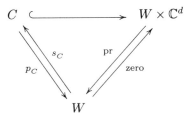

If $f_1, \ldots, f_d$ is a regular sequence, it follows from the preceding lemma that $C = W \times \mathbb{C}^d$. In general, since C is defined by a homogeneous ideal, it is a subcone of $W \times \mathbb{C}^d$, i.e., C is invariant under multiplication by $\mathbb{C}^*$ on the fibres $\mathbb{C}^d$.

In spite of the marvelous brevity of this algebraic definition of normal cone, its geometry is not so simple. Considerable study, beginning with [**32**], has been devoted to the case where W is a nonsingular subvariety. For example, if $W = P$ is a point, then $C_P V$ is the *tangent cone* to V at P; if V is a hypersurface in $\mathbb{C}^d$, and P is the origin, one may check that $C_P V$ is the hypersurface in $\mathbb{C}^d$ defined by the leading homogeneous term of an equation for V. However, as is evident from the preceding section, the normal cones of interest for intersection theory are usually defined by ideals that are not prime ideals, i.e. W is a subscheme of V, but not usually a subvariety. There has been extensive recent study of associated graded rings $\bigoplus I^t/I^{t+1}$ in commutative algebra; one hopes that useful criteria for identifying the irreducible components of C, with their multiplicities, may emerge.

The *projective normal cone* $\mathbb{P}(C) = \mathbb{P}(C_W V)$ is defined by

$$\mathbb{P}(C) = \mathrm{Proj}\Big(\bigoplus I^t/I^{t+1}\Big).$$

In concrete terms, if generators for I are chosen as above, $\mathbb{P}(C)$ is the subscheme of $W \times \mathbb{P}^{d-1}$ defined by the same equations that define C in $W \times \mathbb{C}^d$.

A closely related and equally important construction is that of the *blow-up* of a variety V along a subscheme W. This is a variety $\widetilde{V} = \mathrm{Bl}_W V$, together with a proper morphism $\pi \colon \widetilde{V} \longrightarrow V$ satisfying:

(i) The inverse image scheme $E = \pi^{-1}(W)$ is a Cartier divisor on $\widetilde{V}$, called the *exceptional divisor*: at each $Q \in E$, the ideal $I\mathcal{O}_{Q,\widetilde{V}}$ has one generator.

(ii) E is isomorphic to $\mathbb{P}(C_W V)$, and the mapping from E to W induced by π is the projection from $\mathbb{P}(C)$ to W:

$$\begin{array}{ccc} \mathbb{P}(C) = E & \hookrightarrow & \widetilde{V} = \mathrm{Bl}_W V \\ \downarrow & & \downarrow \pi \\ W & \hookrightarrow & V \end{array}$$

(iii) The induced mapping from $\widetilde{V} \smallsetminus E$ to $V \smallsetminus W$ is an isomorphism.

A quick definition of $\mathrm{Bl}_W V$ is

$$\mathrm{Bl}_W V = \mathrm{Proj}\Big(\bigoplus_{t=0}^{\infty} I^t\Big),$$

the mapping π determined by the isomorphism of A with I^0. If $f_1, \ldots, f_d$ generate I, $\mathrm{Bl}_W V$ is the subvariety of $V \times \mathbb{P}^{d-1}$ defined by the kernel of the canonical homomorphism from $A[X_1, \ldots, X_d]$ onto $\bigoplus I^t$. In case $f_1, \ldots, f_d$ is a regular sequence, $\mathrm{Bl}_W V$ is defined by the equations $f_i X_j - f_j X_i$, $i < j$, by the lemma of §2.4. In general, $\mathrm{Bl}_W V$ is the closure of the graph of the morphism from $V \smallsetminus W$ to $\mathbb{P}^{d-1}$ defined by $(f_1 : \ldots : f_d)$.

Note that, since A is a domain, $\bigoplus I^t$ is also a domain, so $\mathrm{Bl}_W V$ is a variety. The identification of $E = \pi^{-1}(W)$ with $\mathbb{P}(C)$ follows from the canonical isomorphism

$$(\bigoplus I^t) \otimes_A A/I = \bigoplus I^t/I^{t+1}.$$

Over the subvariety of $\mathbb{P}^{d-1}$ where the coordinate X_i is not zero, E is defined by the equation f_i, since $f_j = (X_j/X_i) f_i$.

One important consequence of this construction is that each irreducible component of $E = \mathbb{P}(C)$ has dimension $d-1$. Indeed, E is locally defined by one equation in the d-dimensional variety $\widetilde{V}$, and any such subscheme has pure codimension one.

The above constructions globalize to the case of an arbitrary proper closed subscheme W of an arbitrary variety V. If $\mathfrak{J}$ is the ideal sheaf of W in V, they are written

$$C_W V = \mathrm{Spec}(\bigoplus \mathfrak{J}^t/\mathfrak{J}^{t+1}),$$
$$\mathbb{P}(C_W V) = \mathrm{Proj}(\bigoplus \mathfrak{J}^t/\mathfrak{J}^{t+1}),$$
$$\mathrm{Bl}_W V = \mathrm{Proj}(\bigoplus \mathfrak{J}^t).$$

They may be constructed by covering V by affine neighborhoods, over which the preceding constructions apply, and gluing over the overlaps.

In case the embedding of W in V is a *regular embedding*, i.e., local equations for the ideal of W in V form a regular sequence in local rings of V, then $C_W V$ is a vector bundle, called the *normal bundle* to W in V, and also denoted $N_W V$. If V and W are nonsingular, this agrees with the definition of $N_W V$ as the quotient of tangent bundles:

$$0 \longrightarrow T_W \longrightarrow T_V|_W \longrightarrow N_W V \longrightarrow 0$$

When D is an effective Cartier divisor, on a variety X, $N_D X$ is the restriction to D of the associated line bundle $\mathcal{O}_X(D)$ on X. If $E = \mathbb{P}(C)$ is the exceptional divisor on the blow-up $\widetilde{V}$ of a variety V along a subscheme W, then

$$N_E \widetilde{V} = \mathcal{O}_{\widetilde{V}}(E)|_E = \mathcal{O}_C(-1)$$

is also the dual line bundle to the *canonical line bundle* $\mathcal{O}_C(1)$ on $\mathbb{P}(\mathbb{C})$.

It is a useful exercise to examine a normal cone which is not a vector bundle. For example, if W is the intersection of two curves that have common components in the plane V, then $C_W V$ will have irreducible components that lie over each irreducible component of W, and other varieties as well. If the curves are written in the form $D_1 + E$, $D_2 + E$, where D_1 and D_2 meet properly, then $C_W V$ has components over each component of E and over each point in $D_1 \cap D_2$, including those points that are in E. To see this, let d_1, d_2, e be polynomials in $R = \mathbb{C}[X,Y]$ defining D_1, D_2, E, and set $I = (d_1 e, d_2 e)$, the ideal of W. One verifies that the kernel of the homomorphism $A/I[U_1, U_2] \longrightarrow \bigoplus I^n/I^{n+1}$, which takes U_i to $d_i e \bmod I^2$, is generated by $d_2 U_1 - d_1 U_2$. Therefore $C_W V$ is the subscheme of $W \times \mathbb{C}^2$ defined by $d_2 U_1 - d_1 U_2$, from which one may read off the components of C.

2.6. Deformation to the normal cone

In light of the principle of continuity, a reason why one can expect to use normal cones to compute intersection products is because there is a *deformation* from the given embedding of a subscheme W of V to the zero section embedding of W in the normal cone $C = C_W V$. The affine version of this is a closed embedding

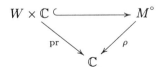

with $M° = M_W° V$ a variety of dimension one greater than $\dim(V)$, such that over $t \neq 0$, the embedding of $W \times \{t\}$ in $M_t°$ is isomorphic to the given embedding of W in V, while the embedding of $W \times \{0\}$ in $M_0°$ is isomorphic to the zero section embedding of W in $C_W V$.

Suppose V is affine with coordinate ring A, and W is defined by the ideal $I = (f_1, \ldots, f_d)$. Let M be the closure of the graph of the morphism

$$(V \smallsetminus W) \times \mathbb{C}^* \longrightarrow \mathbb{P}^d$$

by $(P, t) \longmapsto (f_1(P):\ldots:f_d(P):t)$, in $V \times \mathbb{C} \times \mathbb{P}^d$. Note that $W \times \mathbb{C}$ is embedded in M by

$$W \times \mathbb{C} = W \times \mathbb{C} \times \{(0:\ldots:0:1)\} \subset M.$$

Over $t = 0$, one sees that the fibre M_0 of M contains the blow-up $\widetilde{V} = \mathrm{Bl}_W V$, but this is disjoint from $W \times \{0\}$. We shall see that the complement to $\widetilde{V}$ in M_0 is the normal cone $C = C_W V$, so that $M° = M \smallsetminus \widetilde{V}$ is the desired deformation space.

An algebraic version of this deformation was studied by Gerstenhaber [24] using the graded ring B defined by

$$B = \cdots \oplus I^n T^{-n} \oplus \cdots \oplus I T^{-1} \oplus A \oplus AT \oplus \cdots \oplus AT^n \oplus \cdots,$$

i.e., $B = \bigoplus_{n=-\infty}^{\infty} I^{-n} T^n$, with $I^m = A$ for $m \leq 0$, and T an indeterminate. One may define $M°$ to be the affine variety whose coordinate ring is B. The projection from $M°$ to $\mathbb{C}$ corresponds to the canonical inclusion of $\mathbb{C}[T]$ in B, and the embedding of $W \times \mathbb{C}$ in $M°$ to the canonical surjection of B onto $A/I[T]$. Since the canonical homomorphism from $A[T]$ to B becomes an isomorphism after inverting T:

$$B_T \cong A[T]_T,$$

the embedding $W \times \mathbb{C} \subset M° \smallsetminus \rho^{-1}(0)$ is isomorphic to the trivial embedding of $W \times \mathbb{C}$ in $V \times \mathbb{C}$. Over $T = 0$, since

$$B/TB \cong \bigoplus_{n=0}^{\infty} I^n/I^{n+1},$$

we see that $M_0° = C_W V$, with W embedded as the zero section.

An equivalent construction of this deformation is to define M to be the blow-up of $V \times \mathbb{C}$ along $W \times \{0\}$. The normal cone to this embedding is the cone

$$C \oplus \mathbb{1} = \mathrm{Spec}(\bigoplus_n I^n/I^{n+1} \otimes_{A/I} A/I[T]),$$

so the exceptional divisor is the projective completion $\mathbb{P}(C \oplus 1)$ of C, where $C = C_W V$. The blow-up $\widetilde{V} = \mathrm{Bl}_W V$ is also contained in M as a divisor, and if $\rho\colon M \longrightarrow \mathbb{C}$ is the projection to $\mathbb{C}$, then the scheme $M_0 = \rho^{-1}(0)$ is the sum of two Cartier divisors, $\mathbb{P}(C \oplus 1)$ and $\widetilde{V}$, which meet in $\mathbb{P}(C)$. We have a commutative diagram:

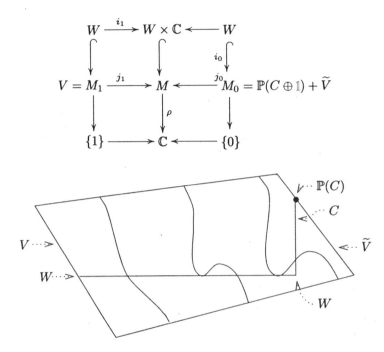

The last construction of the deformation works equally well for arbitrary varieties V and arbitrary closed subschemes W of V.

The deformation to the normal cone often functions as an algebro-geometric analogue of the topologists' tubular neighborhood. Note, however, that even if V and W are nonsingular, there is usually no neighborhood of W in V that is, even complex analytically, isomorphic to a neighborhood in the normal bundle. In the singular case, the normal cone may be a bundle over W even when no neighborhood of W in V is locally a product of W with a disc.

2.7. Intersection products: a preview

Normal cones will be basic to our general construction of intersection products, even when the intersection is not proper. To intersect hypersurfaces (Cartier divisors) $H_1, \ldots, H_d$ on a variety V, consider the set-up:

$$\begin{array}{ccc} H_1 \cap \cdots \cap H_d & \longrightarrow & V \\ \downarrow & & \downarrow \delta \\ H_1 \times \cdots \times H_d & \longrightarrow & V \times \cdots \times V \end{array}$$

2.7. INTERSECTION PRODUCTS: A PREVIEW

To intersect subvarieties $V_1, \ldots, V_s$ of a nonsingular variety X, consider ("reduction to the diagonal"):

$$\begin{array}{ccc} V_1 \cap \cdots \cap V_s & \longrightarrow & V_1 \times \cdots \times V_s \\ \uparrow & & \uparrow \\ X & \xrightarrow{\delta} & X \times \cdots \times X \end{array}$$

Here δ denotes the diagonal embedding.

Each of these is a special case of the situation:

$$\begin{array}{ccc} W & \hookrightarrow & V \\ \uparrow & & \uparrow \\ X & \xrightarrow{f} & Y \end{array}$$

Here V is an n dimensional subvariety of Y, and $f \colon X \longrightarrow Y$ is a *regular embedding* of codimension d, i.e. X is locally defined in Y by a regular sequence of d elements; and W is the intersection scheme $X \cap V$, the subscheme of Y defined locally by the equations for X and for V. Our goal is to construct and compute an *intersection product* $X \bullet V$, which will be an equivalence class of cycles of dimension $n - d$ on X (in fact on W).

Since f is a regular embedding, the normal cone to X in Y is a vector bundle of rank d on X, denoted $N_X Y$. There is a canonical embedding of $C_W V$ in $N_X Y$:

$$\begin{array}{ccc} C_W V & \hookrightarrow & N_X Y \\ \downarrow & & \downarrow \\ W & \hookrightarrow & X \end{array}$$

In fact, if V and Y are affine with coordinate rings A and B, and W and X are defined by ideals I and J, respectively, then $JA = I$, so there is a surjection

$$\bigoplus J^t / J^{t+1} \longrightarrow \bigoplus I^t / I^{t+1},$$

which corresponds to the embedding of $C_W V$ in $N_X Y$.

We saw in the previous section that C has pure dimension $n = \dim(V)$. By the procedure of §2.1, C determines a cycle $[C]$ on $N_X Y$. The class $X \bullet V$ will be constructed by "intersecting the cycle $[C]$ with the zero section of $N_X Y$". Explicitly, $X \bullet V$ will be represented by a cycle of the form $\sum n_i [V_i]$, where V_i are $(n-d)$-dimensional subvarieties of X and $[C]$ is (rationally) equivalent to the pull-back cycle $\sum n_i [N_X Y |_{V_i}]$. (The next few sections will explain these terms.)

In fact if $[C] = \sum m_i [C_i]$, with C_i the irreducible components of C and Z_i is the support of the cone C_i, then the intersection of C_i with the zero section is a well-defined cycle class α_i on Z_i, and

$$X \bullet V = \sum m_i \alpha_i.$$

We shall see that in the case of hypersurfaces considered in §1.5, these Z_i are the distinguished varieties found by Severi and Lazarsfeld, and the contribution $i(Z_i)$ is simply the degree of $m_i \alpha_i$.

In case intersection is *proper*, i.e. $\dim W = n - d$, then $X \bullet V$ is a well-defined cycle on W. If $W_1, \ldots, W_r$ are the irreducible components of W, then

$$[C] = \sum m_i [N_X Y|_{W_i}],$$

so $X \bullet V = \sum m_i [W_i]$. The coefficients m_i agree with the multiplicities discussed in §2.4 in case $n = d$.

In the opposite extreme, when $V = X$, so $W = X$, then $C_W V = X$ is the zero section of $N_X Y$. The intersection of the zero section with itself will be the "top Chern class" of $N_X Y$, and we will have the self-intersection formula

$$X \bullet X = c_d(N_X Y) \cap [X].$$

CHAPTER 3

Divisors and Rational Equivalence

3.1. Homology and cohomology

Before beginning to develop a theory of rational equivalence, Chern classes, intersection products, etc., let us look to topology for a model.

A complex projective variety X has *homology* groups $H_q X$ and *cohomology* groups $H^p X$. Homology is covariant, cohomology contravariant. There are *cup products*
$$H^p X \otimes H^q X \xrightarrow{\cup} H^{p+q} X$$
and *cap products*
$$H^p X \otimes H_q X \xrightarrow{\cap} H_{q-p} X,$$
which make $H^* X = \bigoplus H^p X$ a skew-commutative graded ring, and $H_* X = \bigoplus H_q X$ an $H^* X$-module. If $f : X \longrightarrow Y$ is a mapping, $f^* : H^* Y \longrightarrow H^* X$ is a homomorphism of rings, and one has the important *projection formula*,
$$f_*(f^* a \cap b) = a \cap f_* b,$$
for $a \in H^* Y$, $b \in H_* X$.

Any k-dimensional subvariety V of X determines a class denoted $\mathrm{cl}(V)$ in $H_{2k} X$. This can be seen from the fact that V can be triangulated to be an oriented $2k$-circuit. If $f : X \longrightarrow Y$ is a morphism, and $f(V) = W$, then
$$f_* \mathrm{cl}(V) = \deg(V/W) \mathrm{cl}(W),$$
where $\deg(V/W)$ is defined as follows. If $\dim(W) < \dim(V)$, then $\deg(V/W) = 0$. If $\dim(W) = \dim(V)$, there is an open $W^\circ \subset W$ such that $V \cap f^{-1}(W^\circ) \longrightarrow W^\circ$ is a finite-sheeted topological covering, and $\deg(V/W)$ is the number of sheets of this covering; algebraically, $\deg(V/W)$ is the degree of the function field $R(V)$ of V as a field extension of $R(W)$.

If X is nonsingular of dimension n, capping with $\mathrm{cl}(X)$ gives the *Poincaré duality* isomorphism
$$H^p X \xrightarrow[\cong]{\cap \mathrm{cl}(X)} H_{2n-p} X.$$

In the nonsingular case, $H_* X \cong H^* X$ has an intersection product. If A and B are subvarieties of X of dimensions a and b, one may therefore define $A \bullet B$ in $H_{2a+2b-2n}(X)$. This product may be refined by using relative groups:
$$\mathrm{cl}(A) \in H_{2a}(A) \cong H^{2n-2a}(X, X \smallsetminus A)$$
and similarly $\mathrm{cl}(B) \in H^{2n-2b}(X, X \smallsetminus B)$, so
$$\mathrm{cl}(A) \cup \mathrm{cl}(B) \in H^{4n-2a-2b}(X, X \smallsetminus (A \cap B))$$
$$\cong H_{2a+2b-2n}(A \cap B),$$

and thus $A \bullet B$ lives in $H_{2a+2b-2n}(A \cap B)$. Note that if the intersection is proper, this last group is free on the classes of irreducible components of $A \cap B$, so this determines $A \bullet B$ as a cycle; in other words, this gives a topological construction of intersection multiplicities.

Complex vector bundles E in X have *Chern classes* $c_i(E)$ in $H^{2i}X$ satisfying:
  (i) $c_0(E) = 1$; $c_i(E) = 0$ if $i > \operatorname{rank}(E)$.
  (ii) If $f\colon Y \longrightarrow X$, $c_i(f^*E) = f^*c_i(E)$.
  (iii) If $0 \longrightarrow E' \longrightarrow E \longrightarrow E'' \longrightarrow 0$ is an exact sequence, then
$$c_i(E) = \sum_{j+k=i} c_j(E') \cup c_k(E'').$$

  (iv) If L is an algebraic line bundle on X and s is a section of L with zero-scheme $D_s \neq X$, then
$$c_1(L) \cap \operatorname{cl}(X) = \operatorname{cl}([D_s]),$$
where $\operatorname{cl}([D_s]) = \sum m_i \operatorname{cl}(D_i)$ if $[D_s] = \sum m_i D_i$ is the cycle of D_s.

For a line bundle $L \in H^1(X, \mathcal{O}_X^*)$, $c_1(L) \in H^2(X, \mathbb{Z})$ may be constructed as the coboundary of L from the exact sequence
$$0 \longrightarrow \mathbb{Z} \xrightarrow{2\pi i} \mathcal{O}_X \xrightarrow{\exp} \mathcal{O}_X^* \longrightarrow 0.$$

For a vector bundle E of rank r, let $\mathbb{P}(E)$ be the projective bundle of lines in E, with projection $p\colon \mathbb{P}(E) \longrightarrow X$, and universal (tautological) exact sequence
$$0 \longrightarrow L_E \longrightarrow p^*E \longrightarrow Q_E \longrightarrow 0$$
with L_E a line bundle. The line bundle $\mathcal{O}_E(1)$ is defined to be the *dual* of L_E. Let $\zeta = c_1(\mathcal{O}(1))$. Following Grothendieck [28], one may then define the Chern classes of E by the identity
$$(*) \qquad \zeta^r + p^*c_1(E)\zeta^{r-1} + \cdots + p^*c_r(E) = 0.$$
(Such an equation exists by the structure of $H^*\mathbb{P}(E)$ as an H^*X-algebra.)

The *total Chern class* $c(E)$ is defined to be $1 + c_1(E) + \cdots + c_r(E)$. The *total Segre class* of E is the formal inverse:
$$s(E) = c(E)^{-1},$$
so $s_0(E) = 1$, $s_1(E) = -c_1(E)$, $s_2(E) = c_1(E)^2 - c_2(E)$, etc. A calculation using $(*)$ shows that
$$s_i(E) \cap \operatorname{cl}(X) = p_*\bigl(\zeta^{r-1+i} \cap \operatorname{cl}(\mathbb{P}(E))\bigr).$$

This gives an alternate construction of Chern classes—or at least of their images in homology: define $s_i(E) \cap \operatorname{cl}(X)$ by this last formula, and invert formally to obtain $c(E) \cap \operatorname{cl}(X)$.

For complete (compact) varieties, singular homology is satisfactory. To extend to arbitrary varieties, *Borel-Moore* homology, constructed from locally finite chains, is more appropriate. With this homology, every variety V has a fundamental homology class $\operatorname{cl}(V)$. [2]

3.2. Divisors

A *Weil divisor* on an n-dimensional variety X is an $(n-1)$-cycle on X, i.e. a finite formal combination $\sum n_i [V_i]$ of subvarieties of codimension 1. A *Cartier divisor* on X is determined by local data consisting of a covering $\{U_i\}$ of X, and rational functions $f_i \in R(U_i)^* = R(X)^*$, such that on overlaps $U_i \cap U_j$, f_i/f_j is a nowhere vanishing regular function, i.e., $f_i/f_j \in \Gamma(U_i \cap U_j, \mathcal{O}^*)$. Local data $\{U_i', f_i'\}$ define the same Cartier divisor if $f_i/f_j' \in \Gamma(U_i \cap U_j', \mathcal{O}^*)$ for all i, j.

If D is a Cartier divisor on X given by local data $\{U_i, f_i\}$, and V is any subvariety of X, then the functions f_i for $U_i \cap V \neq \emptyset$ are unique up to units in the local ring $\mathcal{O}_{V,X}$. Such a rational function f is called a *local equation* for D at V. If V is of codimension one in X, and we write $f = a/b$ with $a, b \in \mathcal{O}_{V,X}$, we may define the *order* of D at V by

$$\text{ord}_V(D) = \text{ord}_V(f) = \ell(\mathcal{O}_{V,X}/(a)) - \ell(\mathcal{O}_{V,X}/(b)),$$

where ℓ denotes the length; note that since $\mathcal{O}_{V,X}$ has dimension one, $\mathcal{O}_{V,X}/(a)$ and $\mathcal{O}_{V,X}/(b)$ have dimension zero, so finite length. It is not hard to verify that this is independent of the choice of a and b.

Each Cartier divisor D on X determines an associated Weil divisor, denoted $[D]$, by

$$[D] = \sum \text{ord}_V(D)[V],$$

the sum over the codimension one subvarieties V of X. The Cartier divisors on X form a group $\text{Div}(X)$, the sum $D + E$ of two Cartier divisors being defined by multiplying local equations for D and E. The mapping $D \mapsto [D]$ defines a homomorphism

$$\text{Div}(X) \longrightarrow Z_{n-1}(X)$$

from $\text{Div}(X)$ to the group $Z_{n-1}(X)$ of Weil divisors on X.

A Cartier divisor D on X determines a *line bundle* $\mathcal{O}(D) = \mathcal{O}_X(D)$. If $\{U_i, f_i\}$ are local data for D, the transition functions for $\mathcal{O}(D)$ from the U_j neighborhood to the U_i neighborhood are the units f_i/f_j; thus a section of $\mathcal{O}(D)$ is given by a collection of regular functions s_i on U_i such that

$$s_i = (f_i/f_j) \cdot s_j$$

on $U_i \cap U_j$. A Cartier divisor D is *effective* if it is defined by local equations f_i which are regular; in this case $s_i = f_i$ determines a canonical section of $\mathcal{O}(D)$. Equivalently an effective Cartier divisor is a subscheme of X whose ideal is locally defined by one equation; this subscheme is the zero-scheme of the canonical section of $\mathcal{O}(D)$.

Any $f \in R(X)^*$ defines a *principal* Cartier divisor $\text{div}(f)$. Two Cartier divisors D and E determine isomorphic line bundles on X if and only if they differ by a principal divisor. It is not hard to show that any line bundle on a variety comes from some divisor, so

$$\text{Div}(X)/\text{Principal divisors} \cong \text{Pic}(X).$$

Note that in topology a Cartier divisor D determines a *cohomology* class $c_1(\mathcal{O}(D))$ in $H^2 X$, while $[D]$ determines a *homology* class $\text{cl}[D]$ in $H_{2n-2}X$. One can show that

$$c_1(\mathcal{O}(D)) \cap \text{cl}(X) = \text{cl}[D].$$

We will develop a rational equivalence theory with analogous properties. For this last formula to hold, note that it is necessary that the class of $[\operatorname{div}(f)]$ must be zero for any principal divisor $\operatorname{div}(f)$.

3.3. Rational equivalence

For any variety (or scheme) X over any field K, let $Z_k X$ be the group of k-cycles $\sum n_i[V_i]$ on X, i.e. the free abelian group on the k-dimensional subvarieties of X. Two k-cycles are *rationally equivalent* if they differ by a sum of cycles of the form
$$\sum [\operatorname{div}(f_i)],$$
where $f_i \in R(W_i)^*$, with W_i subvarieties of X of dimension $k+1$. (Strictly speaking, $[\operatorname{div}(f_i)]$ was defined in the preceding section to be a k-cycle on W_i; we freely use the same notation for the cycles they define on any larger variety.) The group of k-cycles modulo rational equivalence on X is denoted $A_k X$, and we write
$$A_* X = \bigoplus A_k X = Z_* X / \sim,$$
where $\sim$ denotes rational equivalence.

Although the preceding definition is usually simplest to work with, it may be shown to be equivalent to the following more classical one. Two k-cycles are rationally equivalent if they differ by a sum
$$\sum n_i \big([V_i(0)] - [V_i(\infty)]\big)$$
with n_i integers, V_i subvarieties of $X \times \mathbb{P}^1$ whose projections to $\mathbb{P}^1$ are dominant, and $V_i(0)$ and $V_i(\infty)$ the scheme-theoretic fibres of V_i over 0 and ∞, regarded as subschemes of $X = X \times \{0\}$ and $X \times \{\infty\}$.

Note that $A_n X = \mathbb{Z}[X] = \mathbb{Z}$ if X is an n-dimensional variety. More generally, if X is a scheme of dimension n, then $A_n X$ is the free abelian group on the n-dimensional irreducible components of X.

If $f\colon X \longrightarrow Y$ is a proper morphism, the formula
$$f_*[V] = \deg(V/f(V))[f(V)]$$
determines a homomorphism $f_*\colon Z_k X \longrightarrow Z_k Y$. To have a covariant ("homology") theory, the following fact is basic:

THEOREM. *If $f\colon X \longrightarrow Y$ is proper, and α and α' are rationally equivalent cycles on X, then $f_*\alpha$ and $f_*\alpha'$ are rationally equivalent cycles on Y.*

Thus there is an induced homomorphism, the *push-forward*
$$f_*\colon A_k X \longrightarrow A_k Y,$$
making A_k a covariant functor for proper morphisms.

For example, if X is a projective curve, and $Y = \operatorname{Spec}(K)$ is a point, the theorem asserts the familiar fact that, for a rational function r on X,
$$\sum \operatorname{ord}_P(r)[R(P):K] = 0,$$
i.e. r "has as many zeros as poles". Another important case is when X and Y are n-dimensional varieties, and f is surjective: if $r \in R(X)^*$, then
$$f_*[\operatorname{div}(r)] = [\operatorname{div}(N(r))].$$

Here $N(r) \in R(Y)^*$ is the norm of r, the determinant of multiplication by r on the finite-dimensional $R(Y)$-space $R(X)$. This formula is a consequence of the basic lemma in §1.6. The theorem, in fact, can be deduced from these two cases, cf. [**16**, §1.4].

In particular, if X is complete, i.e. the projection $p\colon X \longrightarrow \mathrm{Spec}(K)$ is proper, the degree of a zero cycle is well defined on rational equivalence classes. We set

$$\int_X \alpha = \deg(\alpha) = p_*(\alpha),$$

identifying $A_0(\mathrm{Spec}(K))$ with $\mathbb{Z}$.

There is an important class of morphisms $f\colon X \longrightarrow Y$ for which there is a contravariant *pull-back* $f^*\colon A_k Y \longrightarrow A_{k+n} X$, where n is the relative dimension of f. For any k-dimensional subvariety V of Y, $f^{-1}(V)$ will be a subscheme of X of pure dimension $k+n$, and we will define

$$f^*[V] = [f^{-1}(V)].$$

(Note that $f^{-1}(V)$ denotes the inverse image scheme, defined by pulling back equations for V in Y; its cycle is defined as in §2.1.) This class of morphisms includes:

(1) projections $p\colon Y \times T \longrightarrow Y$, T an n-dimensional variety; here $p^*[V] = [V \times T]$;

(2) projections $p\colon E \longrightarrow Y$ (resp. $\mathbb{P}(E) \longrightarrow Y$) from a bundle to its base; here $p^*[V] = [E|_V]$ (resp. $[\mathbb{P}(E|_V)]$);

(3) open embeddings $j\colon U \longrightarrow Y$, with $n=0$, and $j^*[V] = [V \cap U]$. If U is the complement of a closed subscheme X of Y, and i is the inclusion of X in Y, the sequences

$$A_k X \xrightarrow{i_*} A_k Y \xrightarrow{j^*} A_k U \longrightarrow 0$$

are exact;

(4) any dominant (nonconstant) morphism from an $(n+1)$-dimensional variety to a nonsingular curve.

A class of mappings including these for which this pull-back is well defined on rational equivalence classes is the class of *flat* morphisms; $f\colon X \longrightarrow Y$ is flat if each local ring $\mathcal{O}_{V,X}$ is flat as a module over $\mathcal{O}_{W,Y}$, with $W = \overline{f(V)}$. This includes all smooth morphisms. For most applications here, the above examples suffice.

The following proposition is needed to complete the construction of intersection product outlined in §2.7.[3]

PROPOSITION. *Let E be a vector bundle of rank r on X, $p_E\colon E \longrightarrow X$ the projection. Then the pull-back homomorphisms*

$$p_E^*\colon A_k X \longrightarrow A_{k+r} E$$

are all isomorphisms.

If $s_E\colon X \longrightarrow E$ is the zero section embedding, and α is any k-cycle or cycle class on E, define the *intersection of α by the zero section*, denoted $s_E^*(\alpha)$, to be the class in $A_{k-r}(E)$ that pulls back to α:

$$p_E^*(s_E^*(\alpha)) = \alpha.$$

Note that by the proposition, α is equivalent to a cycle of the form $\sum n_i [E|_{V_i}]$, and clearly the intersection of such a cycle with the zero section should be $\sum n_i [V_i]$.

In particular, in the situation of §2.7, the intersection class $X \bullet V$ is a well-defined class in $A_{n-d}(W)$, with $W = X \cap V$. Indeed, the normal cone C to W in V determines an n-cycle $[C]$ on the restriction N of $N_X Y$ to W, and we may set

$$X \bullet V = s_N^*[C].$$

As for the proof of the proposition, the surjectivity of p_E^* follows by a Noetherian induction argument, using the exact sequence of (3) above. The injectivity, and in fact a formula for the inverse s_E^*, uses Chern classes (§4).

Another elementary operation on rational equivalence is the *exterior product*

$$A_k X \otimes A_\ell Y \xrightarrow{\times} A_{k+\ell}(X \times Y)$$

defined by $[V] \times [W] = [V \times W]$.

3.4. Intersecting with divisors

If D is a Cartier divisor on X, and α a k-cycle on X, we define an intersection class

$$D \bullet \alpha \in A_{k-1}(Z),$$

where Z is the intersection of the support of D (the union of varieties at which local equations are not units) and the support of α (the union of varieties appearing in α with nonzero coefficients). By linearity it suffices to define $D \bullet [V]$ if V is a subvariety of X. Let i be the inclusion of V in X. There are two cases:

(i) If V is not contained in the support of D, then by restricting local equations, D determines a Cartier divisor, denoted i^*D, on V. In this case, set

$$D \bullet [V] = [i^*D],$$

the associated Weil divisor of i^*D on V. In this case $D \bullet [V]$ is a well-defined cycle.

(ii) If $V \subset \mathrm{Supp}(D)$, then the line bundle $\mathcal{O}_X(D)$ restricts to a line bundle $i^*\mathcal{O}_X(D)$ on V. Choose a Cartier divisor C on V whose line bundle is isomorphic to this line bundle: $\mathcal{O}_V(C) \cong i^*\mathcal{O}_X(D)$, and set

$$D \bullet [V] = [C],$$

the associated Weil divisor of C. Since C is well defined up to a principal divisor on V, $[C]$ is well defined in $A_{k-1}(V)$.

In case D is an effective Cartier divisor on X, this class $D \bullet [V]$ agrees with the class $D \bullet V$ constructed in §2.7. In case (i) this is immediate, while in case (ii) it amounts to the fact that for a Cartier divisor C on a variety V, the cycle of the zero section $[V]$ is rationally equivalent to the cycle $[\pi^{-1}(C)]$ in the line bundle $L = \mathcal{O}_V(C)$, with $\pi \colon L \longrightarrow V$ the projection. When C is effective, corresponding to a section s of L, an explicit rational equivalence may be constructed as follows (cf. §2.6): let

$$Z = \left\{ (P, (\lambda_0 : \lambda_1)) \in L \times \mathbb{P}^1 \,\middle|\, \lambda_0 s\pi(P) = \lambda_1 P \right\}.$$

Then $Z(0) = \pi^{-1}(C)$, and $Z(\infty)$ is the zero section.

In general, because of the ambiguity in case (ii), $D \bullet \alpha$ is only defined up to rational equivalence. If the restriction of the line bundle $\mathcal{O}_X(D)$ to D is trivial, however, $D \bullet \alpha$ can always be defined as a *cycle*. Namely, if $V \subset \mathrm{Supp}(D)$, set $D \bullet [V] = 0$. This applies when D is the fibre of a morphism from X to a nonsingular curve; the cycle $D \bullet \alpha$ is then called the *specialization* of α.

This intersection product satisfies the formal properties one would expect for a "cap product". For example:

(1) If $\alpha \sim \alpha'$, then $D \bullet \alpha = D \bullet \alpha'$ in $A_*(\mathrm{Supp}(D))$.

(2) If $D - D'$ is principal, then $D \bullet \alpha = D' \bullet \alpha$ in $A_*(\mathrm{Supp}(\alpha))$.

(3) (*Projection formula*) If $f \colon Y \longrightarrow X$ is a proper surjective morphism of varieties, D a Cartier divisor on X, and α a k-cycle on Y, then

$$f'_*(f^*D \bullet \alpha) = D \bullet f_*\alpha$$

in $A_{k-1}(Z)$, with $Z = \mathrm{Supp}(D) \cap f(\mathrm{Supp}\,\alpha)$, and $f' \colon f^{-1}(Z) \longrightarrow Z$ the morphism induced by f. There is a similar compatibility with flat pull-backs.

From (1) and (2) it follows that the operation product $D \bullet \alpha$ determines products

$$\mathrm{Pic}(X) \otimes A_k X \longrightarrow A_{k-1}(X).$$

For a line bundle L and cycle class α we shall write $c_1(L) \cap \alpha$ for this product:

$$c_1(\mathcal{O}_X(D)) \cap \alpha = D \bullet \alpha.$$

This will be the basis for the study of Chern classes in the next chapter.

If D is an effective divisor on X and f is the inclusion of D in X, it follows from (1) that $[V] \longmapsto D \bullet [V]$ determines a "Gysin" homomorphism

$$f^* \colon A_k X \longrightarrow A_{k-1} D.$$

This is the key to showing that the general intersection product is well defined on rational equivalence classes. Note that this is a strong form of the principle of continuity: all such intersection operations, applied to rationally equivalent cycles, will give classes of the same degree. It also includes the statement that specialization respects rational equivalence.

In fact properties (2) and (3) are straightforward to prove. Property (1) then follows from (2) and the following basic commutativity law, on which most of the subsequent theory depends.

LEMMA. *Let D and E be Cartier divisors on an n-dimensional variety X. Then*

$$D \bullet [E] = E \bullet [D]$$

in $A_{n-2}(\mathrm{Supp}(D) \cap \mathrm{Supp}(E))$.

Consider the case where X is a surface, and $\pi \colon X \longrightarrow \mathbb{C}^2$ is a proper birational morphism which is an isomorphism except over $(0,0)$, and $Z = \pi^{-1}((0,0))$ is a curve. Let D and E be the inverse images of the two axes $\mathbb{C} \times \{0\}$ and $\{0\} \times \mathbb{C}$. Then $D = D' + D''$ and $E = E' + E''$, where D' and E' map isomorphically to the two axes, while D'' and E'' are supported on Z. In this case $D \bullet [E]$ is the point where E' meets Z, and $E \bullet [D]$ is the point where D' meets Z. These two points may well be different, but one knows they are rationally equivalent, because Z is a *connected* curve, all of whose components are *rational* curves.

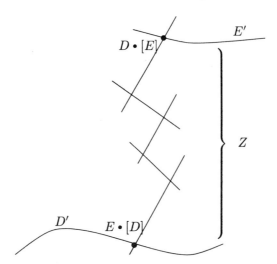

Although one may fashion a proof along these lines (cf. [4]), there is now a simpler proof. Roughly speaking, one blows up X along various subschemes to reduce to the case where D and E are sums and differences of effective divisors D_i and E_j, and such that each intersection of D_i with E_j is either proper (in which case the commutativity is easy) or $D_i = E_j$ (when it is evident). See [**16**, §2.4] for details.

3.5. Applications

Let us apply the preceding results to a situation considered in the first two sections. If $H_1, \ldots, H_d$ are hypersurfaces (effective Cartier divisors) on an n-dimensional variety X, we may define, for any k-cycle α on X, a class

$$H_1 \bullet \ldots \bullet H_d \bullet \alpha \in A_{k-d}(Z),$$

$Z = \bigcap H_i \cap \mathrm{Supp}(\alpha)$. Inductively, this class is defined to be

$$H_1 \bullet (H_2 \bullet \ldots \bullet H_d \bullet \alpha).$$

The commutativity law says that this product is independent of the order of the H_i. If $k = d$, and Z is complete, this class has a well-defined degree, denoted $\int H_1 \bullet \ldots \bullet H_d \bullet \alpha$. When $\alpha = [X]$, we omit it from the notation, and write simply $H_1 \bullet \ldots \bullet H_d$.

Suppose a nonsingular point P on X, rational over the ground field ($R(P) = K$), is an isolated point of intersection of the intersection of n hypersurfaces $H_1, \ldots, H_n$, $n = \dim(X)$. Shrinking X, assume that H_i meet only at P. Let $\pi \colon \widetilde{X} \longrightarrow X$ be the blow-up of X at P, $E = \mathbb{P}^{n-1}$ the exceptional divisor. Then

$$\pi^* H_i = m_i E + G_i,$$

where m_i is the multiplicity of H_i at P; the intersection of G_i with E is the projective tangent cone $\mathbb{P}(C_P H_i)$. We will show that if these projective tangent cones do not meet, then

$$i(P) = i(P, H_1 \bullet \ldots \bullet H_n) = m_1 \cdot \ldots \cdot m_n.$$

Note first that, since local equations for H_i at P form a regular sequence, the intersection product $H_1 \bullet \ldots \bullet H_n$ is the cycle $i(P)[P]$, with $i(P)$ defined as in §1.6

3.5. APPLICATIONS

or §2.4. Let η be the projection from E to P. Then since $\bigcap G_i = \emptyset$,

$$0 = \eta_*(G_1 \bullet \ldots \bullet G_n)$$
$$= \eta_*\big((\pi^*H_1 - m_1 E) \bullet \ldots \bullet (\pi^*H_n - m_n E)\big).$$

Now, by the projection formula, $\eta_*(\pi^*H_1 \bullet \ldots \bullet \pi^*H_n) = H_1 \bullet \ldots \bullet H_n$ and

$$\eta_*(\pi^*H_{i_1} \bullet \ldots \bullet \pi^*H_{i_k} \bullet E^{n-k}) = H_{i_1} \bullet \ldots \bullet H_{i_k} \bullet \eta_*(E^{n-k}).$$

Since the intersection class $E^{n-k} = E \bullet \ldots \bullet E$ is in $A_k(E)$, $\eta_*(E^{n-k}) \in A_k(P) = 0$ if $0 < k < n$. Thus the only terms that survive are

$$0 = H_1 \bullet \ldots \bullet H_n + (-1)^n m_1 \cdot \ldots \cdot m_n \eta_*(E^n).$$

Now the restriction of $\mathcal{O}_{\widetilde{X}}(E)$ to $E = \mathbb{P}^{n-1}$ is the dual of the bundle $\mathcal{O}_{\mathbb{P}^{n-1}}(1)$, so $E \bullet E$ is represented by minus a hyperplane, and E^n is represented by $(-1)^{n-1}$ times a point. Therefore

$$\eta_*(E^n) = (-1)^{n-1}[P],$$

and thus $H_1 \bullet \ldots \bullet H_n = m_1 \cdot \ldots \cdot m_n [P]$.

Let us also reconsider the case where three surfaces H_1, H_2, H_3 in a nonsingular threefold X contain a nonsingular curve C as a (scheme-theoretic) component. Let $\pi: \widetilde{X} \longrightarrow X$ be the blow-up of X along C, and let E be the exceptional divisor, $\eta: E \longrightarrow C$ the projection. Let

$$\pi^*H_i = E + G_i.$$

The hypersurfaces G_i do not meet in E, and the intersection of the H_i *outside* C is represented by the class $\pi_*(G_1 \bullet G_2 \bullet G_3)$. Expanding as above, and noting that again $\eta_*(E) = 0$ for dimension reasons, we obtain

$$\pi_*(G_1 \bullet G_2 \bullet G_3) = H_1 \bullet H_2 \bullet H_3 + \sum H_i \bullet \eta_*(E^2) - \eta_*(E^3).$$

The last two terms therefore determine the contribution of C to the intersection product. The problem is to compute $\eta_*(E^i)$.

We have seen that $E = \mathbb{P}(N)$, where N is the normal bundle to C in X. With this identification, the restriction of $\mathcal{O}_{\widetilde{X}}(E)$ to E (which is the normal bundle to E in $\widetilde{X}$) is the dual of the bundle $\mathcal{O}_N(1)$. Referring to §3.1, one expects the formulas

$$\eta_*(E^2) = \eta_*\big(-c_1(\mathcal{O}_N(1)) \cap [\mathbb{P}(N)]\big) = -[C],$$
$$\eta_*(E^3) = \eta_*\big(c_1(\mathcal{O}_N(1))^2 \cap [\mathbb{P}(N)]\big) = -c_1(N) \cap [C],$$

where $c_1(N)$ is the *first Chern class* of the bundle N. In the next section we will develop the necessary theory of Chern classes for rational equivalence.

Combining these results, it follows that the contribution of C to the intersection product $H_1 \bullet H_2 \bullet H_3$ is

$$\big(\sum H_i\big) \bullet C - c_1(N) \cap [C].$$

For example, if $X = \mathbb{P}^3$, and C is a complete intersection of surfaces of degrees a and b, then $N = \mathcal{O}(a) \oplus \mathcal{O}(b)$, and $\deg(C) = ab$, so the degree of $c_1(N) \cap [C]$ is $(a+b)(ab)$, and the total contribution of C to the Bézout number is

$$ab\big(\sum \deg(H_i) - (a+b)\big),$$

as found by Salmon (§1.3). With the machinery of Chern classes, one can also compute the contribution when C is not a complete intersection.

It is similarly possible to work out the contribution of the Veronese V in the intersection of five hypersurfaces H_i representing conics tangent to five fixed general conics (§1.4). Blowing up $\mathbb{P}^5$ along V as above one has

$$\pi^* H_i = 2E + G_i,$$

where $G_1, \ldots, G_5$ are hypersurfaces that do not meet in the exceptional divisor E. Knowing the Chern classes of the tangent bundles to $V \cong \mathbb{P}^2$ and $\mathbb{P}^5$, one knows the Segre classes of the normal bundle $N_V \mathbb{P}^5$, and it is a pleasant exercise (see [**22**]) to verify that

$$\int G_1 \bullet \ldots \bullet G_5 = 3264.$$

This approach to excess intersection problems was developed primarily by B. Segre [**56**], with related work by Severi and Todd. All were searching for constructions which would yield *invariants* of varieties, generalizing the notion of genus for curves. For a subvariety V of a variety X, let $\tilde{X}$ be the blow-up of X along V, E the exceptional divisor, η the projection from E to V. The classes $\eta_*(E^i)$ were called the *covariants* of the embedding of V in X. We shall see that, up to sign, they are the *inverse Chern classes* of the normal bundle of V in X, at least when V and X are nonsingular. For example, Segre constructed the *canonical* classes of a nonsingular variety V by applying this construction to the diagonal embedding of V in $X = V \times V$; the formal inverse of $\sum \eta_*(E^i)$ on $V \times V$ projects to the total Chern class of T_V on V.

CHAPTER 4

Chern Classes and Segre Classes

4.1. Chern classes of vector bundles

Eventually one wants a contravariant "cohomology" theory A^*X to go with the covariant theory A_*X, and Chern classes of vector bundles on X should lie in A^*X. Although such theories exist, at this time there is not yet a simple geometric construction of such a cohomology theory. Indeed, it would be extremely useful to have such a theory, perhaps analogous to Goresky's realization of ordinary cohomology via "geometric cocycles" [26].

At any rate, any such theory should have "cap products"

$$A^i X \otimes A_k X \xrightarrow{\cap} A_{k-i}X,$$

and Chern classes $c_i(E) \in A^i X$. In particular, a bundle E on X should determine homomorphisms

$$A_k X \xrightarrow{c_i(E) \cap \underline{}} A_{k-i} X,$$

by $\alpha \longmapsto c_i(E) \cap \alpha$. In this section we construct such Chern class operations directly. They will satisfy properties expected from topology (§3.1).

For a line bundle L on a variety (or scheme) X, to define $c_1(L) \cap \alpha$ it suffices to define $c_1(L) \cap [V]$, for V a subvariety of X. Choose a Cartier divisor C on V such that the restriction $L|_V$ of L to V is isomorphic to $\mathcal{O}_V(C)$, and set

$$c_1(L) \cap [V] \doteq [C].$$

Note that if $L = \mathcal{O}_X(D)$, then $c_1(L) \cap \alpha = D \bullet \alpha$, the intersection product of §3.4. It follows from the discussion of §3.4 that this operation respects rational equivalence classes, and satisfies the expected formal properties. For example, there is a *projection formula*

$$f_*(c_1(f^*L) \cap \alpha) = c_1(L) \cap f_*\alpha$$

for $f: Y \longrightarrow X$ proper, L a line bundle on X, α a cycle class on Y. Similarly we have a *commutativity* property

$$c_1(M) \cap (c_1(L) \cap \alpha) = c_1(L) \cap (c_1(M) \cap \alpha)$$

for line bundles L, M on X, $\alpha \in A_*X$. Thus any polynomial in first Chern classes of line bundles on X—or on any variety that X maps to—operates on A_*X. In addition there are the elementary formulas:

$$c_1(L \otimes M) \cap \alpha = c_1(L) \cap \alpha + c_1(M) \cap \alpha,$$
$$c_1(L^{-1}) \cap \alpha = -c_1(L) \cap \alpha.$$

Now if E is a vector bundle of rank r on X, define *Segre class* operators $s_i(E)$,

$$s_i(E) \cap \underline{}: A_k X \longrightarrow A_{k-i}X,$$

as follows. Let $p\colon \mathbb{P}(E) \longrightarrow X$ be the projective bundle of E, $\mathcal{O}_E(1)$ the canonical line bundle on $\mathbb{P}(E)$, and for α in $A_k X$ set

$$s_i(E) \cap \alpha = p_*\bigl(c_1(\mathcal{O}_E(1))^{r-1+i} \cap p^*\alpha\bigr).$$

One shows easily that $s_i(E) = 0$ for $i < 0$, and that $s_0(E) = 1$ (i.e. $s_0(E) \cap \alpha = \alpha$ for all α). Basic properties such as projection formulas and commutativity of these classes follow readily from corresponding formulas for first Chern classes of line bundles.

Now we define *Chern class* operators

$$c_i(E) \cap \underline{} \colon A_k X \longrightarrow A_{k-i} X$$

by formally inverting the Segre classes (cf. §3.1):

$$1 + c_1(E) + c_2(E) + \cdots = (1 + s_1(E) + s_2(E) + \cdots)^{-1},$$

i.e., set $c_0(E) = 1$, and $c_1(E) = -s_1(E)$, $c_2(E) = s_1(E)^2 - s_2(E)$, ... Explicitly:

$$c_p(E) = (-1)^p \det \begin{vmatrix} s_1(E) & 1 & 0 & \cdots & 0 \\ s_2(E) & s_1(E) & 1 & & \vdots \\ \vdots & & & & 0 \\ \vdots & & & s_1(E) & 1 \\ s_p(E) & \cdots & & \cdots & s_1(E) \end{vmatrix}.$$

Properties such as the *projection formula*

$$f_*(c_i(f^*E) \cap \alpha) = c_i(E) \cap f_*(\alpha)$$

and the *commutativity* property

$$c_i(E) \cap (c_j(F) \cap \alpha) = c_j(F) \cap (c_i(E) \cap \alpha)$$

follow formally from the corresponding facts for Segre classes. Less obvious but also true are the *vanishing property*

$$c_i(E) = 0 \quad \text{for } i > \operatorname{rank} E,$$

and the *Whitney sum formula*

$$c_i(E) = \sum_{j+k=i} c_j(E')c_k(E'')$$

for an exact sequence $0 \longrightarrow E' \longrightarrow E \longrightarrow E'' \longrightarrow 0$ of vector bundles. There are also formulas for Chern classes of tensor, exterior or symmetric products.

Although we shall not carry out complete proofs of these statements here, a basic ingredient is a *splitting principle*: an equation among Chern classes of vector bundles in a given relation with each other is true if:

(i) the equation is valid when the bundles each have filtrations by subbundles such that the quotient bundles are line bundles, and

(ii) the given relation is preserved by pull-back.

This principle is a simple consequence of the fact that for any bundle E of rank r on X,

$$p^* \colon A_k X \longrightarrow A_{k+r-1}(\mathbb{P}(E))$$

is *injective*, which follows from the fact that $s_0(E) = 1$. For on $\mathbb{P}(E)$, p^*E contains the universal line bundle L_E, with quotient bundle Q_E: repeating the process on

Q_E yields a composite $f: Y \longrightarrow X$ of projective bundles, so f^*E is filtered, and f^* injects A_*X in A_*Y.

If E is filtered, with line bundle quotients $L_1, \ldots, L_r$, the vanishing and Whitney formulas reduce to showing that $c_i(E)$ is the ith elementary symmetric function of $c_1(L_1), \ldots, c_1(L_r)$. For this, one first verifies directly that $\prod c_1(L_i) = 0$ if E has a nowhere vanishing section; one then may apply this to the bundle $p^*E \otimes L_E^\vee$, which gives

$$\prod_{i=1}^r \bigl(c_1(p^*L_i) + c_1(\mathcal{O}_E(1))\bigr) = 0,$$

from which the assertion follows easily.

As usual we define *total* Segre and Chern classes by

$$s(E) = 1 + s_1(E) + s_2(E) + \cdots,$$
$$c(E) = 1 + c_1(E) + c_2(E) + \cdots.$$

These Chern classes may be used to prove the isomorphisms $A_{k-r}X \xrightarrow{\sim} A_k E$ for a vector bundle E of rank r on a variety or scheme X (§3.2). One proves first the isomorphisms

$$\bigoplus_{i=0}^{r-1} A_{k-r+i+1}(X) \longrightarrow A_k \mathbb{P}(E),$$

which take $\sum \alpha_i$ to $\sum c_1(\mathcal{O}_E(1))^i \cap p^*\alpha_i$, p the projection from $\mathbb{P}(E)$ to X. The surjectivity of this mapping is proved by a Noetherian induction as in the affine bundle case; injectivity follows by applying operators $p_*(c_1(\mathcal{O}_E(1))^j \cap \_\_)$, using the identities that $s_i(E) = 0$ if $i < 0$, and $s_0(E) = 1$.

The *projective completion* $\mathbb{P}(E \oplus 1)$ contains E as an open subvariety, complementary to the *hyperplane at infinity* $\mathbb{P}(E)$. From the above isomorphism and the exact sequence

$$A_k \mathbb{P}(E) \longrightarrow A_k \mathbb{P}(E \oplus 1) \longrightarrow A_k E \longrightarrow 0$$

the injectivity of $A_{k-r}X \longrightarrow A_k E$ follows easily. In addition, one derives a formula for the inverse isomorphism $s_E^* \colon A_k E \longrightarrow A_{k-r}X$. Given a subvariety V of E, let $\overline{V}$ be its closure in $\mathbb{P}(E \oplus 1)$. Then,

$$s_E^*([V]) = q_*(c_r(Q) \cap [\overline{V}]);$$

q is the projection from $\mathbb{P}(E \oplus 1)$ to X, and Q is the universal rank r quotient bundle of $q^*(E \oplus 1)$ on $\mathbb{P}(E \oplus 1)$. (Note that Q has a canonical section which vanishes precisely on the zero section of X in E; multiplying by the top Chern class should therefore correspond to intersecting with the zero section.)

4.2. Segre classes of cones and subvarieties

Let W be a subvariety of a variety V. If V and W are nonsingular, or, more generally, if the embedding of W in V is a regular embedding, one has a normal bundle $N_W V$, and one may construct invariants of the embedding by using Chern and Segre classes

$$c_i(N_W V) \cap [W] \qquad \text{and} \qquad s_i(N_W V) \cap [W].$$

In the general case, however, one has only a normal cone $C = C_W V$. We shall see that, although one does not have a general Chern class formalism for cones, there is a useful notion of Segre class.

We shall define a *total Segre class*
$$s(W,V) \in A_*W$$
for any closed subscheme W of a variety V. If $W = V$, set $s(W,V) = [V]$. Otherwise, let $\widetilde{V}$ be the blow-up of V along W, let $E = \mathbb{P}(C)$ be the exceptional divisor, and let $\eta\colon E \longrightarrow W$ be the projection. The i-fold self-intersections $E^i = E \bullet \ldots \bullet E$ of the divisor E are well defined classes in $A_{k-i}(E)$, $k = \dim(V) = \dim(\widetilde{V})$, by the construction of §3.4. We set
$$s(W,V) = \sum_{i \geq 1} (-1)^{i-1} \eta_*(E^i).$$
At least in the nonsingular case, the images of these self-intersections E^i were basic for Segre's construction of invariants [56].

Identifying E with $\mathbb{P}(C)$, the restriction of $\mathcal{O}_{\widetilde{V}}(E)$ to E is the dual of the universal line bundle $\mathcal{O}_C(1)$ on $\mathbb{P}(C)$. Hence $E^i = (-1)^{i-1} c_1(\mathcal{O}_C(1))^{i-1} \cap [\mathbb{P}(C)]$, and therefore
$$s(W,V) = \sum_{i \geq 0} \eta_*\big(c_1(\mathcal{O}_C(1))^i \cap [\mathbb{P}(C)]\big).$$
Note that this last expression makes sense for any cone C on a scheme W; under the assumption that, for each irreducible component C' of C, $\mathbb{P}(C')$ is not empty, we define the Segre class $s(C)$ of the cone C by this formula:
$$s(C) = \eta_*\big(c(\mathcal{O}_C(-1))^{-1} \cap [\mathbb{P}(C)]\big).$$
For an arbitrary cone C, the cone $C \oplus \mathbb{1}$ satisfies this assumption, and one may define $s(C)$ to be $s(C \oplus \mathbb{1})$.

Since the restriction of $\mathcal{O}_{\widetilde{V}}(E)$ to E is also the normal bundle to E in $\widetilde{V}$, another definition of the Segre class is
$$s(W,V) = \eta_*(s(E,\widetilde{V})).$$
This is a special case of the following important formula.

PROPOSITION. *Let $\pi\colon V' \longrightarrow V$ be a proper surjective morphism of varieties, of degree d. Let W be a subscheme of V, $W' = \pi^{-1}(W)$, and let $\eta\colon W' \longrightarrow W$ be the induced morphism. Then*
$$\eta_*(s(W',V')) = d \cdot s(W,V).$$

This is easily proved by blowing up to reduce to the case where W and W' are Cartier divisors, in which case it follows from the formula $f_*[W'] = d[W]$.

When $d = 1$, the proposition expresses the *birational invariance* of Segre classes. When the embedding of W' in V' is regular, e.g. if V' and W' are nonsingular, it gives a formula for $s(W,V)$ in terms of Chern classes of the normal bundle of W' in V'. When all four varieties are nonsingular, it gives a remarkable relation among the Chern classes of the normal bundles; when these Chern classes are known, it can even be used to compute the degree d.

If Z is an irreducible component of W, the coefficient of $[Z]$ in the class $s(W,V)$ is the *multiplicity of V along W at Z*, and is denoted $(e_W V)_Z$. If A is the local ring of V along Z, and I the ideal of W, one may show that
$$\text{length}(A/I^t) = (e_W V)_Z (t^n/n!) + \text{lower terms}$$

for $t \gg 0$ and $n = \text{codim}(W, V)$. In other words, this multiplicity agrees with Samuel's multiplicity for the primary ideal I in the local ring A. If $Z = W$, we write simply $e_W V$. We shall see that other terms in the Segre classes also appear in intersection formulas.

It is illuminating to apply these ideas to verify the *Riemann-Kempf formula*. Fixing a base point on a nonsingular curve C determines morphisms $u_d \colon C^{(d)} \longrightarrow J$ from symmetric products to the Jacobian of C. If W_d is the image of $C^{(d)}$, and $1 \leq d \leq g$, and $D \in C^{(d)}$ is a divisor, the Riemann-Kempf formula states that the multiplicity of W_d at the point $u_d(D)$ is $\binom{g-d+r}{r}$, where g is the genus of C, and r is the dimension of the linear series $|D|$ of D. Indeed, one knows that $|D|$ is the fibre $u_d^{-1}(u_d(D))$, and one may calculate that

$$s(|D|, C^{(d)}) = (1+h)^{g-d+r} \cap [\,|D|\,],$$

where $h = c_1(\mathcal{O}(1))$ on $|D| = \mathbb{P}^r$. Since u_d maps $C^{(d)}$ birationally onto W_d, the proposition applies, giving the multiplicity as[4]

$$\int_{\mathbb{P}^r} (1+h)^{g-d+r} = \binom{g-d+r}{r}.$$

Segre classes will appear frequently in these notes. In the study of holonomic $\mathcal{D}$-modules, important invariants are constructed by intersecting characteristic varieties in cotangent bundles with the zero section; as we shall see, all such intersections can be expressed in terms of Segre classes.

It may be pointed out that, for a vector bundle E, some authors' $s_i(E)$ correspond to our $s_i(E^\vee) = (-1)^i s_i(E)$. The necessity of enlarging their scope to include general cones dictates our convention.

4.3. Intersection forumulas

Recall the situation of the basic construction of intersection products (§2.7):

$$\begin{array}{ccc} W & \xrightarrow{j} & V \\ h \downarrow & & \downarrow g \\ X & \xrightarrow{f} & Y \end{array}$$

with $f \colon X \longrightarrow Y$ a regular embedding of codimension d, V an n-dimensional variety, g a closed embedding, $W = X \cap V = g^{-1}(X)$. Let $N = h^* N_X Y$, $C = C_W V$ the normal cone, which is a closed, n-dimensional subscheme of N. We have defined the intersection product

$$X \bullet V = s_N^*[C] \in A_{n-d}(W).$$

Another description of $X \bullet V$ may be derived from the last formula in §4.1. Let Q be the universal rank d quotient bundle on $\mathbb{P}(N \oplus 1)$, and let q be the projection from $\mathbb{P}(N \oplus 1)$ to W. Then

$$X \bullet V = q_*\big(c_d(Q) \cap [\mathbb{P}(C \oplus 1)]\big).$$

We use the notation $\{\alpha\}_k$ for the k-dimensional component of a cycle or class α on a scheme. Using the Whitney and projection formulas, we have

$$X \bullet V = \{q_*(c(Q) \cap [\mathbb{P}(C \oplus 1)])\}_{n-d}$$
$$= \{q_*(c(q^*(N \oplus 1)) \cdot c(\mathcal{O}(-1))^{-1} \cap [\mathbb{P}(C \oplus 1)])\}_{n-d}$$
$$= \{c(N) \cdot q_*(c(\mathcal{O}(-1))^{-1} \cap [\mathbb{P}(C \oplus 1)])\}_{n-d}.$$

This gives a basic

Intersection formula $\qquad X \bullet V = \{c(N) \cap s(W, V)\}_{n-d}.$

Note first that if Z is an irreducible component of W of the expected dimension $n - d$, then the coefficient of $[Z]$ in $X \bullet V$ is just the multiplicity $(e_W V)_Z$ of V along W at Z. In particular, in the case of *proper intersection*, i.e. $\dim W = n - d$, we recover the formula

$$X \bullet V = \sum (e_W V)_Z [Z],$$

the sum over the irreducible components Z of W.

Since $A_* W = \bigoplus A_* W_i$, where the W_i are the connected components of W, one has a corresponding decomposition for $X \bullet V$:

$$X \bullet V = \sum \{c(N) \cap s(W_i, V)\}_{n-d}.$$

(Notation is abused by writing $c(N) \cap s(W_i, V)$ in place of $c(N|_{W_i}) \cap s(W_i, V)$.)

If Y° is open in Y, and X°, V°, and W° are the intersections of Y° with X, V, and W, then the restriction homomorphism from $A_* W$ to $A_* W^\circ$ takes $X \bullet V$ to $X^\circ \bullet V^\circ$. By means of this localizing principle, which follows immediately from the construction, it suffices to consider the case when W is connected; similarly one may discard any closed subvarieties of dimension less than $n - d$, without loss of information.

Consider the case when the embedding of W in V is a regular embedding of codimension d'. In this case the normal cone C is a subbundle of N. The quotient bundle $E = N/C$ is called the *excess bundle*. Since $s(W, V)$ is given by the inverse Chern class of C, we deduce from the Whitney formula and the above intersection formula the

Excess intersection formula $\qquad X \bullet V = c_{d-d'}(E) \cap [W].$

In case $d' = d$, i.e. regular sequences locally defining X in Y remain regular sequences on V, we recover again the formula $X \bullet V = [W]$.

There is a simple but important refinement of these constructions and formulas. The morphism $g: V \longrightarrow Y$ can be an *arbitrary* morphism; it need not be a closed embedding. Defining W to be the inverse image scheme $g^{-1}(X)$, $h: W \longrightarrow X$ the induced morphism, one still has the normal cone $C = C_W V$ embedded in $N = h^* N_X Y$, and $X \bullet V \in A_* W$ can be constructed by intersecting $[C]$ with the zero section in N. The preceding intersection formulas are equally valid in this generality.

Combined with the birational invariance of Segre classes, this allows an imporant reduction procedure. To compute $X \bullet V$, it suffices to find a proper birational $\pi: V' \longrightarrow V$ for which the class $X \bullet V'$ can be computed. For then the class $X \bullet V'$ pushes forward to $X \bullet V$. Indeed, if $W' = \pi^{-1}(W)$, and $\eta: W' \longrightarrow W$ is the morphism induced by π, then

$$X \bullet V = \eta_*(X \bullet V').$$

This follows from the formula $s(W,V) = \eta_*(s(W',V'))$ and the intersection formula. For example, if W' is regularly embedded in V' with codimension d', and with excess normal bundle $E = (h\eta)^* N_X Y / N_{W'} V'$, then

$$X \bullet V = \eta_*(c_{d-d'}(E) \cap [W']).$$

One may always reduce to this case, with $d' = 1$, by taking V' to be the blow-up of V along W. In this way, many difficult problems can be reduced to the case of divisors and Chern classes.

CHAPTER 5

Gysin Maps and Intersection Rings

5.1. Gysin homomorphisms

If $f\colon X \longrightarrow Y$ is a regular embedding of codimension d, we define *Gysin homomorphisms*
$$f^*\colon A_k Y \longrightarrow A_{k-d} X$$
by the formula $f^*(\sum n_i[V_i]) = \sum n_i(X \bullet V_i)$, where $X \bullet V_i \in A_{k-d}X$ is the intersection product constructed in §§3.3 and 4.3. Verdier's proof that this formula respects rational equivalence [59] uses the deformation to the normal bundle to reduce to the known case where $d=1$. It goes as follows. Let $N = N_X Y$, and let $M°$ be the deformation space constructed in §2.6. Let i be the embedding of N in $M°$ (as a Cartier divisor). The complement of N in $M°$ is identified with $Y \times \mathbb{C}$; let j be the inclusion of $Y \times \mathbb{C}$ in $M°$. Consider the diagram:

$$\begin{array}{ccccccc} A_{k+1}N & \xrightarrow{i_*} & A_{k+1}M° & \xrightarrow{j^*} & A_{k+1}(Y \times \mathbb{C}) & \longrightarrow & 0 \\ & & i^* \downarrow & & \cong \uparrow \text{pr}^* & & \\ & & A_k N & \xleftarrow{\quad\sigma\quad} & A_k Y & & \end{array}$$

Here i^* is the Gysin homomorphism defined for divisors in §3.4. The row is exact (§3.3(3)), and $i^* \circ i_* = 0$ because the normal bundle to N in $M°$ is trivial. Hence there is a *specialization* homomorphism σ as indicated, with $\sigma(\alpha) = i^*\beta$ if $j^*\beta = \text{pr}^*\alpha$. For a subvariety V of Y, with $W = V \cap Y$, it follows that
$$\sigma[V] = i^*[M_W° V] = [C_W V],$$
where $C_W V$ is the normal cone to W in V. One deduces that f^* is the composite
$$A_k Y \xrightarrow{\sigma} A_k N \xrightarrow{s_N^*} A_{k-d}X,$$
which is evidently well defined on rational equivalence classes.

There is a useful strengthening of these Gysin homomorphisms. If $f\colon X \longrightarrow Y$ is a regular embedding of codimension d, and $g\colon Y' \longrightarrow Y$ is an arbitrary morphism, form the fibre square

$$\begin{array}{ccc} X' & \xrightarrow{f'} & Y' \\ g' \downarrow & & \downarrow g \\ X & \xrightarrow{f} & Y \end{array}$$

i.e. $X' = X \times_Y Y' = g^{-1}(X)$. We define *refined Gysin homomorphisms*
$$f^!\colon A_k Y' \longrightarrow A_{k-d}X'$$

by the same formula $f^![V] = X \cdot V$. (This intersection product $X \cdot V$ was constructed in $A_{k-d}(V \cap X')$ at the end of §4.3; as usual we use the same notation for its image in $A_{k-d}X'$.) Similar reasoning shows that $f^!$ is well defined on rational equivalence classes.

The main compatibilities of these Gysin homomorphisms are stated in the following theorems.

THEOREM 1. *Consider a fibre square*

$$\begin{array}{ccc} X' & \xrightarrow{f'} & Y' \\ {\scriptstyle g'}\downarrow & & \downarrow{\scriptstyle g} \\ X & \xrightarrow{f} & Y \end{array}$$

with f a regular embedding of codimension d. (a) If g is proper, and $\alpha \in A_k Y'$, then

$$f^* g_* \alpha = g'_* f^! \alpha \quad \text{in } A_{k-d}X.$$

(b) *If g is flat of relative dimension n, and $\alpha \in A_k Y$, then*

$$g'^* f^* \alpha = f^! g^* \alpha \quad \text{in } A_{k+n-d}X'.$$

(c) *If f' is also a regular embedding of codimension d', set $E = g'^* N_X Y / N_{X'} Y'$. Then, for $\alpha \in A_k Y'$,*

$$f^! \alpha = c_{d-d'}(E) \cap f'^* \alpha \quad \text{in } A_{k-d}X'.$$

(d) *If g is also a regular embedding of codimension e, and $\alpha \in A_k Y$, then*

$$g^! f^* \alpha = f^! g^* \alpha \quad \text{in } A_{k-d-e}X'.$$

(e) *If F is a vector bundle on Y', then for all $\alpha \in A_k Y'$, and all i,*

$$f^!(c_i(F) \cap \alpha) = c_i(f'^* F) \cap f^! \alpha \quad \text{in } A_{k-d-i}X'.$$

For example, if f is proper, then (a) and (c) yield the *self-intersection formula*: for $\alpha \in A_k X$,

$$f^* f_* \alpha = c_d(N_X Y) \cap \alpha \quad \text{in } A_{k-d}X.$$

Note that if $d' = d$ in case (c), the assertion is that $f^! \alpha = f'^* \alpha$.

The proofs of (a)–(e) follow quite easily from facts we have discussed before: (a) from the proposition in §4.2; (b) from an analogous formula for pull-backs of Segre classes by flat morphisms; (c) as in the excess intersection formula (§4.3); (d) is reduced, as in the discussion at the end of §4.3, to the case of divisors, which is the main lemma of §3.4; a similar reduction is used in (e).

THEOREM 2. *Let $f: X \longrightarrow Y$ and $g: Y \longrightarrow Z$ be regular embeddings of codimensions d and e. Then the composite gf is a regular embedding of codimension $d + e$, and if $\alpha \in A_k Z$, then*

$$(gf)^* \alpha = f^*(g^* \alpha) \quad \text{in } A_{k-d-e}X.$$

The equation $(gf)^* = f^* g^*$ also holds when f is a regular embedding and either (i) g and gf are flat, or (ii) gf is a regular embedding and g is flat.

For example, if $p: E \longrightarrow X$ is a vector bundle of rank r, and $s: X \longrightarrow E$ is a section, it follows from (ii) that $s^*: A_k E \longrightarrow A_{k-r}X$ is the inverse isomorphism to p^*; in particular, s^* is independent of the choice of s.

The theorem and the variations stated after it are also valid for the refined Gysin morphisms. If $h\colon Z' \longrightarrow Z$ is any morphism, and $\alpha \in A_*Z'$, then

$$(gf)^!\alpha = f^!(g^!\alpha) \quad \text{in } A_*(X'),$$

with $X' = X \times_Z Z'$. The theorem is straightforward when Z is a vector bundle over Y, and g is the zero section. The general case is reduced to this by a deformation to the normal bundle, cf. [**59**]. We refer to [**16**, §§6, 17] for the general statements and complete proofs.[5]

5.2. The intersection ring of a nonsingular variety

If X is an n-dimensional nonsingular variety (i.e., smooth over the base field), then the diagonal embedding δ of X in $X \times X$ is a regular embedding of codimension n. Given $\alpha \in A_aX$ and $\beta \in A_bX$, a *product* $\alpha \bullet \beta \in A_mX$, $m = a+b-n$, is defined by

$$\alpha \bullet \beta = \delta^*(\alpha \times \beta).$$

Thus the product on A_*X is the composite

$$A_aX \otimes A_bX \longrightarrow A_{a+b}(X \times X) \xrightarrow{\delta^*} A_mX.$$

Note that if $\alpha \in A_aV$ and $\beta \in A_bW$, with V, W closed subschemes of X, then the product $\alpha \bullet \beta$ has a natural well-defined refinement in $A_m(V \cap W)$, namely $\delta^!(\alpha \times \beta)$, with $\delta^!$ the refined Gysin homomorphism constructed from the fibre square:

$$\begin{array}{ccc} V \cap W & \longrightarrow & V \times W \\ \downarrow & & \downarrow \\ X & \xrightarrow{\delta} & X \times X \end{array}$$

All the formulas of this section are valid for such refinements, but for simplicity we write them only in the absolute case.

Define A^pX to be $A_{n-p}X$. Then the product is a homomorphism

$$A^pX \otimes A^qX \xrightarrow{\bullet} A^{p+q}X.$$

Let $1 \in A^0X$ correspond to $[X] \in A_nX$.

If $f\colon Y \longrightarrow X$ is a morphism of nonsingular varieties, then the graph morphism

$$\gamma_f\colon Y \longrightarrow Y \times X$$

is a regular embedding of codimension $n = \dim(X)$. Define $f^*\colon A^pX \longrightarrow A^pY$ by the formula

$$f^*\alpha = \gamma_f^*(\alpha \times [X]).$$

THEOREM. *For X nonsingular, the above product makes A^*X into an associative, commutative ring with unit 1. For a morphism $f\colon Y \longrightarrow X$ of nonsingular varieties, the homomorphism $f^*\colon A^*X \longrightarrow A^*Y$ is a ring homomorphism. If also $g\colon Z \longrightarrow Y$, with Z nonsingular, then $(fg)^* = g^*f^*$.*

The theorem follows quite readily from the general properties of intersection products summarized in §5.1. For example, to prove the associativity of the product, consider the fibre square:

$$\begin{array}{ccc} X & \xrightarrow{\delta} & X \times X \\ \delta \downarrow & & \downarrow \delta \times 1 \\ X \times X & \xrightarrow{1 \times \delta} & X \times X \times X \end{array}$$

Given cycles α, β, γ on X, the equality $\alpha \bullet (\beta \bullet \gamma) = (\alpha \bullet \beta) \bullet \gamma$ is equivalent to the formula

$$\delta^*(1 \times \delta)^*(\alpha \times \beta \times \gamma) = \delta^*(\delta \times 1)^*(\alpha \times \beta \times \gamma).$$

This follows either from Theorem 1(d), (c), or from Theorem 2. We refer to [**16**, §8] for details and refinements.

The formula for f^* also makes sense when $f \colon Y \longrightarrow X$ is any morphism, with X nonsingular. More generally, one may construct "cap products"

$$A^p X \otimes A_q Y \longrightarrow A_{q-p} Y$$

by defining $f^*\alpha \cap \beta$, or $\beta \bullet_f \alpha$, to be $\gamma_f^*(\beta \times \alpha)$. This makes $A_* Y$ into a module over $A^* X$, and one has the *projection formula* $f_*(f^*\alpha \cap \beta) = f_*(\alpha) \cap \beta$, or

$$f_*(\beta \bullet_f \alpha) = f_*(\alpha) \bullet \beta$$

in case f is proper.

If Y is nonsingular, and a subvariety X of Y is regularly embedded in Y by an inclusion i, then for any cycle α on Y,

$$\alpha \bullet [X] = i_* i^*(\alpha) \quad \text{in } A^* Y.$$

More generally, $\alpha \bullet [X] = i^!(\alpha) \in A_*(X \cap \mathrm{Supp}(\alpha))$. The commutativity property (Theorem 1(d)) is used to prove this.

Since the seminar of Chevalley [**10**], the intersection ring $A^* X$ has been known as the *Chow ring* of X. The construction in that seminar was for nonsingular quasiprojective varieties over algebraically closed fields, and was based on a "moving lemma". Chow's work in turn was inspired by ideas and constructions of Severi, many of whose papers were devoted to intersection theory. B. Segre, Todd, van der Waerden, Weil, and Samuel were among the others who studied rings of equivalence classes of cycles. One feature of the present approach, following [**21**], is the elimination of any need for a moving lemma. Our approach is closest to that advocated by B. Segre [**56**]; related ideas have been proposed by many others, including Murre, Mumford, Jouanolou, King, Lascu, Scott, and Gillet.

If V and W are subvarieties of a nonsingular X, the refined intersection class $[V] \bullet [W] = \delta^![V \times W]$ is in $A_m(V \cap W)$, $m = \dim V + \dim W - \dim X$. In particular, any *proper* m-dimensional component Z of $V \cap W$ appears in $[V] \bullet [W]$ with a positive coefficient, the *intersection multiplicity* $i(Z, V \bullet W; X)$. Basic properties of this multiplicity, such as associativity, follow from the refined versions of the theorems in §5.1. If V and W meet transversally along a nonempty open subvariety of Z, it follows from our construction that $i(Z, V \bullet W; X) = 1$. The converse is also true. For this criterion of multiplicity one we refer to [**34**] and [**16**] for algebraic and geometric proofs.

Although several of the above-mentioned authors indicated that some of their constructions made sense on singular varieties, the attempt to bring singular varieties into the general picture was apparently diverted by the notion that it should be possible to intersect general cycles on a singular variety if *rational* coefficients are used. This is possible on normal surfaces and on quotients of nonsingular varieties by finite groups. For example, the intersection of two generating lines on a cone over a plane conic is then one-half the vertex. But, as Zobel [62] points out, this is not possible in general. If $X \subset \mathbb{P}^4$ is the cone over a quadric surface Q, any two lines in Q are rationally equivalent in X, since they are rationally equivalent to generators of the cone. However, the cone over a line in Q meets lines in one family of lines in Q transversally, but is disjoint from lines in the other family. It is interesting that this same cone is used in the example of Dutta, Hochster, and McLaughlin [14].

5.3. Grassmannians and flag varieties

In general the computation of the ring A^*X, for a nonsingular projective variety X, is a very difficult problem. One has $A^0 X = \mathbb{Z}$, and $A^1 X = \text{Pic}(X)$, but for $p \geq 2$, there is little general knowledge of $A^p X$. Mumford [45] showed that, for general surfaces, it is impossible to give $A^2 X$ any natural, finite-dimensional, algebraic geometric structure. Collino [12] has calculated A^*X for X a symmetric product of a curve (in terms of the Chow ring of the Jacobian of the curve), and Bloch and Murre [8] have done the same for certain Fano threefolds. Such calculations use all the special geometry of the varieties in question; there are very few general algorithms.[6]

There is an important class of homogeneous varieties, however, for which the groups $A^p X$ are finitely generated, and the rings A^*X known, at least in principle. The group-theoretic approach is probably most satisfactory (cf. [13], [31]), but we will give more classical descriptions.

The Grassmannian $G = G_d(\mathbb{P}^n)$ of d-planes in $\mathbb{P}^n$ is a nonsingular variety of dimension $(d+1)(n-d)$. Fix a flag

$$A_0 \subsetneq A_1 \subsetneq \cdots \subsetneq A_d \subset \mathbb{P}^d$$

of subspaces, with $a_i = \dim A_i$, and set

$$\Omega(A_0, \ldots, A_d) = \{ L \in G \mid \dim L \cap A_i \geq i, \, 0 \leq i \leq d \}.$$

Then $\Omega(A_0, \ldots, A_d)$ is a subvariety of G, called a *Schubert* variety. Its dimension is

$$\sum_{i=0}^{d}(a_i - i) = \sum_{i=0}^{d} a_i - d(d+1)/2.$$

Its class in A_*G depends only on the integers $0 \leq a_0 < \cdots < a_d \leq n$, and is denoted $(a_0, \ldots, a_d)$. A notation better suited to codimensions, and also used by Schubert, is to define, for $n - d \geq \lambda_0 \geq \cdots \geq \lambda_d \geq 0$,

$$\{\lambda_0, \ldots, \lambda_d\} = (a_0, \ldots, a_d) = [\Omega(a_0, \ldots, a_d)],$$

where $a_i = n - d + i - \lambda_i$. Then $\{\lambda_0, \ldots, \lambda_d\}$ is in $A^{|\lambda|}G$, where $|\lambda| = \sum_{i=1}^{d} \lambda_i$. A third common notation is $\sigma_{\lambda_0, \ldots, \lambda_d}$.

In the usual Plücker embedding of $G_d(\mathbb{P}^n)$ in $\mathbb{P}^N$, $N = \binom{n+1}{d+1} - 1$, the Schubert varieties are defined by linear equations. If $e_0, \ldots, e_n$ are points spanning $\mathbb{P}^n$, and A_i is spanned by $e_0, \ldots, e_{a_i}$, the Schubert variety $\Omega = \Omega(A_0, \ldots, A_d)$ has an open subvariety Ω° consisting of those linear spaces L that can be spanned by $v_0, \ldots, v_d$ with v_i in A_i, but v_i not in the span of $e_0, \ldots, e_{a_i-1}$. The "reduced echelon" form of such a basis identifies Ω° with the affine space of dimension $\sum(a_i - i)$. The complement $\Omega \smallsetminus \Omega^\circ$ is a union of smaller Schubert varieties.

An inductive argument, using the exact sequence of §3.3(3), then shows that the classes $(a_0, \ldots, a_d)$ generate A_*G. We shall see that they form a free basis. As a first step toward understanding the ring structure on A^*G, consider the intersection of classes $(a_0, \ldots, a_d)$ and $(b_0, \ldots, b_d)$ of complementary dimension. The *dual* class to $(a_0, \ldots, a_d)$ is the class

$$(n - a_d, n - a_{d-1}, \ldots, n - a_0).$$

One has the basic *duality*:

$$(a_0, \ldots, a_d) \bullet (b_0, \ldots, b_d) = \begin{cases} (0, \ldots, d) & \text{if } (b_0, \ldots, b_d) \text{ is dual} \\ & \text{to } (a_0, \ldots, a_d), \\ 0 & \text{otherwise.} \end{cases}$$

To see this, one may represent $(b_0, \ldots b_d)$ by $\Omega(B_0, \ldots, B_d)$, where B_i is spanned by the last points $e_{n-b_i}, \ldots, e_n$. The Schubert varieties are then seen to meet transversally in one point when the classes are dual; otherwise they are disjoint.

It follows that the Schubert classes form a free basis for A_*G. Moreover, given any k-cycle α on G, its expression in terms of this basis is described as follows. For each class $(b_0, \ldots, b_d)$ of codimension k, set

$$\alpha_{b_0,\ldots,b_d} = \int_G \alpha \bullet (b_0, \ldots, b_d).$$

Then $\alpha = \sum \alpha_{n-a_d,\ldots,n-a_0} (a_0, \ldots, a_d)$.

One may use this principle to calculate general products. The reader is invited to work out $A^*G_1(\mathbb{P}^3)$ this way. In this case Schubert used a special notation: $1 = (2,3)$, $g = (1,3)$, $g_p = (0,3)$, $g_e = (1,2)$, $g_s = (0,2)$, and $G = (0,1)$ form a basis, and

$$g^2 = g_p + g_e, \quad g \cdot g_p = g \cdot g_e = g_s,$$
$$g_p^2 = g_e^2 = G, \quad g_p \cdot g_e = 0, \quad g \cdot g_s = G.$$

To see the geometry behind the first equation, note that g^2 is represented by the variety of lines in space meeting two general lines. Moving the lines so that they meet, this variety degenerates to the union of the variety of lines through the point of intersection and the variety of lines in the plane of the two lines. Such arguments, standard in classical enumerative geometry, must be fortified with a verification of the multiplicities of intersection; for this one may intersect both sides with a dual basis. Other techniques will be discussed in the next chapter.

With this one may calculate that $g^4 = 2G$: there are two lines meeting four given lines in general position. If C is an irreducible curve of degree d in $\mathbb{P}^3$, and

$$V_C = \{\ell \in G_1(\mathbb{P}^3) \mid \ell \text{ meets } C\},$$

then $[V_C] = d \cdot g$. To verify this, one checks that $[V_C] \bullet g_s = d$, since there are d lines through a general point in a general plane that meet C. It follows that there are $2 \prod \deg(C_i)$ lines meeting four curves $C_1, \ldots, C_4$ in general position. One may similarly count the number of common chords to two space curves, and many other similar numbers.

From the fact that the projective linear group acts transitively on $G_d(\mathbb{P}^n)$, one may deduce that, after putting varieties in general position via translations by this group, all intersection will be transversal, so that the naïve geometric number agrees with the intersection-theoretic multiplicity, at least in characteristic zero [**37**].

There is a similar description for a general *flag manifold*, dating from Ehresmann [**15**]. For $0 \leq d_1 < d_2 < \cdots < d_r < n$, let $F = F(d_1, \ldots, d_r; n)$ denote the flag manifold whose points are flags of subspaces

$$L_1 \subset \cdots \subset L_r \subset \mathbb{P}^n$$

with $\dim L_i = d_i$. Fix $e_0, \ldots, e_n$ spanning $\mathbb{P}^n$ as before. The Schubert varieties in F are described by an array with r rows

$$\begin{pmatrix} a_0 & a_1 & \cdots & a_{d_1} & & & \\ b_0 & b_1 & \cdots & & \cdots & b_{d_2} & \\ \vdots & \vdots & & & & & \\ c_0 & c_1 & \cdots & & & \cdots & c_{d_r} \end{pmatrix}$$

where each row is an increasing sequence of integers between 0 and n, and each row is a subset of the next. The Schubert variety consists of all flags $L_1 \subset \cdots \subset L_r$ such that L_i satisfies the Schubert condition prescribed by the ith row, with respect to the standard flag. The dimension of this variety is

$$\sum(a_i - i) + {\sum}'(b_i - i) + \cdots + {\sum}'(c_i - i),$$

where the primes denote that only those terms not counted in the preceding row are included. The classes of these cycles form a basis for $A_*(F)$, and the dual of such a class is obtained by replacing each row by the dual Schubert condition.

5.4. Enumerating tangents

Let $I = F(0, d; n)$ be the *incidence variety* of points on d-planes in $\mathbb{P}^n$. Then $A_*(I)$ has a basis of classes of the form $(a_0, \ldots, \overset{*}{a_k}, \ldots, a_d)$. Here $(a_0, \ldots, a_d)$ is a Schubert condition for d-planes; if $A_0 \subset \cdots \subset A_d$ is a fixed flag, with $\dim A_i = a_i$, then $(a_0, \ldots, \overset{*}{a_k}, \ldots, a_d)$ is the class of the variety

$$\left\{ (P, L) \in I \mid \dim L \cap A_i \geq i,\ 0 \leq i \leq d,\text{ and } P \in A_k \right\},$$

whose dimension is $\sum(a_i - i) + k$. The dual class is

$$(n - a_d, \ldots, n \overset{*}{-} a_k, \ldots, n - a_0).$$

Let V be a subvariety of $\mathbb{P}^n$ of codimension $e \leq d+1$. Let $V' \subset I$ be the closure of

$$\left\{ (P, L) \in I \mid P \in V_{\text{reg}},\ \dim(L \cap T_P V) \geq d - e + 1 \right\}.$$

Here V_{reg} is the nonsingular locus of V, and $T_P V$ is the tangent $(n-e)$-plane to V at P. Then V' is a subvariety of I of codimension $d + 1$, which measures the pointed d-planes that *touch* V. For many enumerative problems involving tangents, it suffices to compute the class $[V']$ in $A^{d+1}(I)$.

If M is a linear subspace of codimension $d - k + 1$, then the class of M' is one of the basic Schubert classes, which we denote μ_k:

$$\mu_k = \{1, \ldots, \overset{*}{1}, 0, \ldots, 0\}$$
$$= (n - d - 1, n - d, \ldots, n - d \overset{*}{+} k - 1, n - d + k + 1, \ldots, n - 1, n).$$

By calculating intersections with dual classes, one verifies that
$$[V'] = m_{d-e+1}\mu_0 + m_{d-e}\mu_1 + \cdots + m_0\mu_{d-e+1},$$
where m_i is the ith *class* of V; namely m_i is the degree of the closure of
$$\left\{ P \in V_{\text{reg}} \,\middle|\, \dim T_P V \cap A \geq i - 1 \right\},$$
where A is a general $(e+i-2)$-plane.

For concreteness, consider the case where $d=1$, $n=2$, and $C=V$ is a plane curve. Then
$$[C'] = n\nu + m\mu,$$
where $n = m_0$ is the degree of C, $m = m_1$ the class of C (§1.2), and
$$\nu = \mu_1 = \left[\{(P, \ell) \,|\, \ell \text{ is a fixed line}\}\right],$$
$$\mu = \mu_0 = \left[\{(P, \ell) \,|\, P \text{ is a fixed point}\}\right].$$

With this, one may calculate the number of curves in a given r-parameter family of curves which are tangent to r given curves in general position. Let $\mathcal{C} = \{C_t\}_{t \in T}$ be an r-dimensional family of plane curves. The *characteristics* $\mu^i \nu^{r-i}$ of the family are the numbers
$$\mu^i \nu^{r-i} = \#\{t \,|\, C_t \text{ passes though } i \text{ general points}$$
$$\text{and is tangent to } r - i \text{ general lines}\}.$$

Given r curves $C_1, \ldots, C_r$ in general position, let $n_i = \deg(C_i)$, $m_i = \text{class}(C_i)$. Then the number of curves in the family tangent to $C_1, \ldots, C_r$ is
$$\prod_{i=1}^{r} (m_i \mu + n_i \nu).$$
This is evaluated by expanding formally, and substituting the characteristics for each $\mu^i \nu^{r-i}$.

For example, if $\mathcal{C}$ is the family of all plane conics, then $\mu^5 = 1$, $\mu^4 \nu = 2$, and $\mu^3 \nu^2 = 4$, as one sees by the fact that the condition to be tangent to a line is a quadric in $\mathbb{P}^5$. For the others one has the Veronese as an excess component (cf. §1.4), by one may conclude by the duality of conics that $\mu^i \nu^j = \mu^j \nu^i$; so $\mu^2 \nu^3 = 4$, $\mu \nu^4 = 2$, $\nu^5 = 1$. Thus if $C_1, \ldots, C_5$ are conics, one finds
$$(2\mu + 2\nu)^5 = 3264$$
conics tangent to five given conics in general position.

It should be pointed out that computation of charcteristics can be very difficult. For the family of all plane curves of degree ≥ 5, apparently no one has even guessed what the answers should be.[7]

On the other hand, the above tangency formula is easy to prove, including the generalization to arbitrary dimensions. Let $\mathcal{C}(r)$ be the closure in $I \times \cdots \times I \times T$ (with r copies of I) of the set
$$\{(P_1, \ell_1) \times \cdots \times (P_r, \ell_r) \times t \,|\, \text{each } P_i \text{ is simple on } C_t$$
$$\text{and } \ell_i \text{ is the tangent line to } C_t \text{ at } P_i\}.$$

Consider the projection
$$f \colon \mathcal{C}(r) \longrightarrow I \times \cdots \times I \qquad (r \text{ copies}).$$

5.4. ENUMERATING TANGENTS

Using transversality, one sees that the desired number is the degree of the intersection class

$$[\mathcal{C}(r)] \bullet_f ([C'_1] \times \cdots \times [C'_r])$$

constructed as in §5.2. Writing out the classes $[C'_i] = m_i\mu + n_i\nu$, the conclusion follows. Note that, by transversality, any lower-dimensional subset of T may be discarded or added without changing these numbers; in particular, one may take T to be a projective variety.

One may realize the equation $[C'] = m\mu + n\nu$ geometrically by deforming C to an n-fold line ℓ via projection from a general point Q:

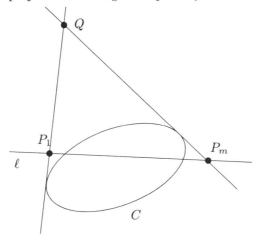

The condition to be tangent to C deforms to n times the condition to be tangent to ℓ, plus the sum of the conditions to pass through the points P_i where tangents from Q to C meet ℓ. With this approach the intersection theory can be carried out on the original parameter space. The essential point is that, for generic such deformations, the contribution of the "Veronese" of multiple curves remains constant: no solutions enter or leave this locus of degenerate solutions at either end of the deformation. For details and other approaches see [16] and [18].

It may be pointed out that the basis for $A_*(I)$ used here and in other enumerative problems is *not* the basis one obtains by realizing I as a projective bundle over $G_d(\mathbb{P}^n)$, cf. §4.1. The notation for this basis follows Martinelli [43].

CHAPTER 6

Degeneracy Loci

6.1. A degeneracy class

Let $\sigma\colon E \longrightarrow F$ be a homomorphism of vector bundles of ranks e and f on an n-dimensional variety X. For $k \leq \min(e, f)$, set

$$D_k(\sigma) = \{\, x \in X \mid \mathrm{rank}(\sigma(x)) \leq k \,\}.$$

This degeneracy locus has a natural structure as a closed subscheme of X, locally defined by the vanishing of $(k+1)$-minors of a matrix representation of σ. One expects $D_k(\sigma)$ to be m dimensional, where

$$m = n - (e-k)(f-k),$$

but in general one can only state that each irreducible component of $D_k(\sigma)$ has dimension at least m. Our object is to construct a class

$$\mathbf{D}_k(\sigma) \in A_m(D_k(\sigma)),$$

to give a formula for the image of $\mathbf{D}_k(\sigma)$ in $A_m X$ in terms of Chern classes of E and F, and to investigate when $\mathbf{D}_k(\sigma)$ is determined by the scheme $D_k(\sigma)$.

To construct $\mathbf{D}_k(\sigma)$, let $d = e - k$, and let $G = G_d(E)$ be the *Grassmannian bundle* of d-planes in E, with projection $\pi\colon G \longrightarrow X$. On G one has a universal exact sequence

$$0 \longrightarrow S \longrightarrow E_G \longrightarrow Q \longrightarrow 0$$

with $\mathrm{rank}\, S = d$, $\mathrm{rank}\, Q = k$ and $E_G = \pi^* E$. The composite $S \longrightarrow E_G \xrightarrow{\sigma} F_G$ determines a section, denoted s_σ, of the bundle $S^\vee \otimes F_G$. The zero scheme $Z(s_\sigma)$ of this section projects onto $D_k(\sigma)$; let

$$\eta\colon Z(s_\sigma) \longrightarrow D_k(\sigma)$$

be the morphism induced by π. If s_0 is the zero section embedding of G in $S^\vee \otimes F_G$, one has a fibre square:

$$\begin{array}{ccc} Z(s_\sigma) & \longrightarrow & G \\ \downarrow & & \downarrow {s_0} \\ G & \xrightarrow{s_\sigma} & S^\vee \otimes F_G \end{array}$$

Since s_σ is a regular embedding, we may construct the refined intersection class $s_\sigma^![G] \in A_m(Z(s_\sigma))$; note that $m = \dim(G) - \mathrm{rank}(S^\vee \otimes F_G)$. Set

$$\mathbf{D}_k(\sigma) = \eta_*(s_\sigma^![G]) \in A_m(D_k(\sigma)).$$

Because $\mathbf{D}_k(\sigma)$ is constructed by a succession of our intersection operations, it is compatible with other such operations, e.g. by pull-backs by flat morphisms or regular embeddings. In particular, $\mathbf{D}_k(\sigma)$ may also be constructed by pull-back from a *universal* case. Let

$$H = \mathrm{Hom}(E, F) = E^\vee \otimes F,$$

a bundle over X. Inside H there is a subcone D_k consisting of mappings of rank at most k. Locally D_k is a product of X and the variety of $e \times f$ matrices of rank at most k; the latter variety is known [33] to be a reduced, irreducible Cohen-Macaulay variety of the expected dimension $ef - (e-k)(f-k)$. Giving a morphism $\sigma\colon E \longrightarrow F$ corresponds to giving a section t_σ of H. Then $D_k(\sigma) = t_\sigma^{-1}(D_k)$, and

$$\mathbf{D}_k(\sigma) = t_\sigma^![D_k].$$

Note that the assertions about the dimension of $D_k(\sigma)$ follow from this statement. In addition, it follows that

$$\mathbf{D}_k(\sigma) = [D_k(\sigma)]$$

precisely when $\mathrm{depth}(D_k(\sigma), X) = \mathrm{codim}(D_k(\sigma), X) = (e-k)(f-k)$; this means that for all $x \in D_k(\sigma)$, the ideal of $D_k(\sigma)$ in $\mathcal{O}_{x,X}$ contains a regular sequence of length $(e-k)(f-k)$. If X is Cohen-Macaulay, e.g. nonsingular, this is equivalent to $D_k(\sigma)$ having the expected codimension. Without this depth condition, even if $D_k(\sigma)$ has the right codimension, $\mathbf{D}_k(\sigma)$ will be a cycle whose support is $D_k(\sigma)$ but whose coefficients are smaller than those in $[D_k(\sigma)]$.

It remains to compute the image of $\mathbf{D}_k(\sigma)$ in $A_m X$. By the theory of §5, one has

$$s_\sigma^![G] = c_{\mathrm{top}}(S^\vee \otimes F_G) \cap [G]$$

in $A_m G$. The required class is then the image of this class in $A_m X$. The answer, to be verified in the next section, is the *Giambelli-Thom-Porteous formula*:

$$\mathbf{D}_k(\sigma) = \Delta_{f-k}^{(e-k)}(c(F-E)) \cap [X].$$

Here $\Delta_q^{(p)}(c)$ denotes the determinant of the p by p matrix

$$\begin{pmatrix} c_q & c_{q+1} & \cdots & c_{q+p-1} \\ c_{q-1} & c_q & & c_{q+p-2} \\ \vdots & & & \vdots \\ c_{q-p+1} & \cdots\cdots\cdots & & c_q \end{pmatrix}$$

and $c(F-E) = c(F)/c(E) = c(F) \cdot s(E)$.

This formula yields a geometric construction for the Chern classes $c_i(F)$ of a bundle F of rank f. Let $e = f - i + 1$, and let E be the trivial bundle of rank e. Then $\sigma\colon E \longrightarrow F$ is given by e sections $s_1, \ldots, s_e$ of F, and $D_{f-i}(\sigma)$ by the locus where these sections become dependent. Then $\mathbf{D}_{f-i}(\sigma)$ is a class in $A_{n-i}(D_{f-i}(\sigma))$ that represents $c_i(F) \cap [X]$, since $\Delta_i^{(1)}(c(F-E)) = c_i(F)$. If F is generated by its sections, and $s_1, \ldots, s_e$ are chosen generically, then

$$\mathbf{D}_{f-i}(\sigma) = [D_{f-i}(\sigma)] = c_i(F) \cap [X].$$

6.2. Schur polynomials

For any formal series $c = 1 + c_1 + c_2 + \cdots$ with c_i in any commutative ring, and any finite sequence $\lambda = (\lambda_1, \ldots, \lambda_d)$ of integers, define $\Delta_\lambda(c)$ to be $\det(c_{\lambda_i+j-i})$, i.e.

$$\Delta_{\lambda_1,\ldots,\lambda_d}(c) = \det \begin{vmatrix} c_{\lambda_1} & c_{\lambda_1+1} & \cdots & c_{\lambda_1+d-1} \\ c_{\lambda_2-1} & c_{\lambda_2} & & \vdots \\ \vdots & & & \vdots \\ c_{\lambda_d-d+1} & \cdots\cdots\cdots\cdots & & c_{\lambda_d} \end{vmatrix}.$$

Note that adding a string a zeros to λ does not change $\Delta_\lambda(c)$. Usually we will assume λ is a *partition*, i.e. $\lambda_1 \geq \lambda_2 \geq \cdots \geq \lambda_d \geq 0$, so λ partitions $|\lambda| = \sum \lambda_i$. Then $\Delta_\lambda(c)$ is the *Schur polynomial* corresponding to λ. If one represents λ by a *Young diagram*

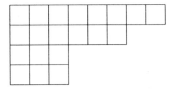

with λ_i boxes in the ith row, the *conjugate* partition is obtained by interchanging rows and columns. A basic formal identity is

(i) $$\Delta_\mu(c) = \Delta_\lambda\left(\left(\sum(-1)^i c_i\right)^{-1}\right)$$

if μ is the conjugate partition to λ. Another is

(ii) $$\Delta_\mu(c) \cdot c_m = \sum \Delta_\mu(c),$$

where the sum is over all $\mu = (\mu_1, \ldots, \mu_{d+1})$ with

$$\mu_1 \geq \lambda_1 \geq \mu_2 \geq \cdots \geq \mu_d \geq \lambda_d \geq \mu_{d+1} \geq 0,$$

and $|\mu| = |\lambda| + m$. More generally there is a Littlewood-Richardson rule for the coefficients $N_{\lambda,\mu,\rho}$ of an arbitrary product [**42**]:

(iii) $$\Delta_\lambda(c) \cdot \Delta_\mu(c) = \sum N_{\lambda,\mu,\rho} \Delta_\rho(c).$$

Some useful formulas for top Chern classes can be expressed in terms of Schur polynomials:

(i) $$c_{\text{top}}(E^\vee \otimes F) = \Delta_{f,\ldots,f}(c(F - E)),$$

where $f = \operatorname{rank} F$, and the subscript is repeated $e = \operatorname{rank} E$ times.

(ii) $$c_{\text{top}}(S^2 E) = 2^e \Delta_{e,e-1,\ldots,1}(c(E)),$$

(iii) $$c_{\text{top}}(\wedge^2 E) = \Delta_{e-1,e-2,\ldots,1}(c(E)).$$

Here $S^2 E$ and $\wedge^2 E$ are symmetric and exterior powers, and $e = \operatorname{rank} E$ [**39**].

We may use (i) to complete the proof of the Giambelli-Thom-Porteous formula. It suffices to verify that, with the notation of §6.1, and any sequence $\lambda = (\lambda_2, \ldots, \lambda_d)$,

$$\pi_*\left(\Delta_\lambda(c(F_G - S)) \cap [G]\right) = \Delta_\mu(c(F - E)) \cap [X],$$

where $\mu = (\lambda_1 - k, \ldots, \lambda_d - k)$.

Note that $\Delta_k(c(F_G-S)) = \Delta_\lambda(c(F_G-E_G)\cdot c(Q))$. Expanding this determinant, one is reduced to showing that, for $\alpha \in A_*X$,

$$\pi_*\big(c_{i_1}(Q)\bullet \ldots \bullet c_{i_d}(Q) \cap \pi^*\alpha\big) = \begin{cases} \alpha & \text{if } i_1 = \cdots = i_d = k, \\ 0 & \text{otherwise.} \end{cases}$$

See [36], [3], or [16] for details.[8]

Schur polynomials have been used by Navarro Aznar [47] to define local invariants of a coherent sheaf $\mathcal{F}$ on a variety X at a point x. Let r be the generic rank of $\mathcal{F}$, and λ a partition with $r \geq \lambda_1 \geq \lambda_2 \geq \cdots$. Shrinking X if necessary, one can find a birational proper map $\pi \colon \widetilde{X} \longrightarrow X$ such that the quotient of $\pi^*\mathcal{F}$ by its torsion subsheaf is the sheaf of sections of a vector bundle E. Define

$$e_x(\mathcal{F}; \lambda) = \int \Delta_\lambda(c(E)) \cap s(\pi^{-1}(x), \widetilde{X}).$$

It follows from the birational invariance of Segre classes (§4.2) that this is independent of choices. When $\mathcal{F}$ is the sheaf of differentials, these classes were studied by Lê and Teissier. MacPherson's *local Euler obstruction* is an alternating sum of some of these invariants.

6.3. The determinantal formula

There is a similar formula for a more general determinantal locus. Let $\sigma \colon E \longrightarrow F$ be a vector bundle homomorphism as before, and let $V_\bullet$ be a flag of subbundles of E:

$$0 \subset V_1 \subset \cdots \subset V_r \subset E.$$

Let $v_i = \operatorname{rank} V_i$, $\lambda_i = f - v_i + i$, $m = \dim(X) - \sum \lambda_i$. Set

$$\Omega(V_\bullet; \sigma) = \big\{ x \in X \mid \dim\big(\operatorname{Ker}(\sigma(x)) \cap V_i(x)\big) \geq i,\ 1 \leq i \leq r \big\}.$$

A similar construction to that in §6.1 constructs a class $\Omega(V_\bullet; \sigma)$ in $A_m(\Omega(V_\bullet; \sigma))$, with analogous properties. The *determinantal formula* states that the image of $\Omega(V_\bullet; \sigma)$ in $A_m X$ is the cap product of

$$\det \begin{vmatrix} c_{\lambda_1}(F-V_1) & c_{\lambda_1+1}(F-V_1) & \cdots & c_{\lambda_1+r-1}(F-V_1) \\ c_{\lambda_2-1}(F-V_2) & c_{\lambda_2}(F-V_2) & & \vdots \\ \vdots & & & \\ c_{\lambda_r-r+1}(F-V_r) & \cdots\cdots\cdots\cdots\cdots & & c_{\lambda_r}(F-V_r) \end{vmatrix}$$

with the fundamental class $[X]$. As in §6.1, the proof of Kempf and Laksov [36] carries over to arbitrary varieties, cf. [16, §14].[8]

This applies to the Grassmannian $X = G = G_d(\mathbb{P}^n) = G_{d+1}(E)$, with E a vector space of dimension $n+1$, and σ the canonical projection from E_G to the universal quotient bundle Q. A flag of subspaces $A_0 \subset \cdots \subset A_d \subset \mathbb{P}^n$ corresponds to a flag $V_0 \subset \cdots \subset V_d \subset E$ of subspaces, with $\dim V_i = \dim A_i + 1$. The Schubert variety is the corresponding degeneracy locus

$$\Omega(A_0, \ldots, A_d) = \Omega(V_\bullet; \sigma).$$

Set $a_i = \dim(A_i)$, $\lambda_i = n - d + i - a_i$. The determinantal formula then yields *Giambelli's formula*:

$$\{\lambda_0, \ldots, \lambda_d\} = (a_0, \ldots, a_d) = \Delta_\lambda(c(Q)).$$

For example, the mth special *Schubert* class
$$\sigma_m = \{m\} = \big[\{\, L \in G \mid L \text{ meets a given } (n-d-m)\text{-plane}\}\big]$$
is equal to $c_m(Q)$, for $m = 1, \ldots, n - d$.

The formula (2) of §6.2 becomes *Pieri's formula*
$$\{\lambda_0, \ldots, \lambda_d\} \cdot \sigma_m = \sum \{\mu_0, \ldots, \mu_d\},$$
the sum over $n - d \geq \mu_0 \geq \lambda_0 \geq \cdots \geq \mu_d \geq \lambda_d \geq 0$ with $\sum \mu_i = \sum \lambda_i + m$. Note that $\{\mu_0, \ldots, \mu_r\} = 0$ if $\mu_0 > n - d$ or $r > d$, corresponding to the facts that $c_i(Q) = 0$ for $i > n - d$ and $s_i(Q^\vee) = c_i(S) = 0$ for $i > d$. Similarly the Littlewood-Richardson rule specializes to a general formula for multiplying Schubert classes.

If X is a nonsingular subvariety of $\mathbb{P}^n = \mathbb{P}(E)$, there is a canonical vector bundle homomorphism
$$\sigma \colon E \otimes \mathcal{O}_X(1) \longrightarrow N_X \mathbb{P}^n$$
which is the composite of the quotient maps $E \otimes \mathcal{O}(1) \longrightarrow T_{\mathbb{P}^n}$ on $\mathbb{P}^n$, and the canonical map $T_{\mathbb{P}^n}|_X \longrightarrow N_X \mathbb{P}^n$. If flags $A_\bullet$, $V_\bullet$ are chosen as above,
$$\Omega(V_\bullet; \sigma) = \{\, x \in X \mid \dim T_x X \cap A_i \geq i,\ 0 \leq i \leq d\,\}.$$
The degree of this locus is the *projective character* $X(a_0, \ldots, a_d)$. The determinantal formula gives a formula for these extrinsic invariants in terms of the intrinsic Chern classes of T_X, and a hyperplane section. The *classes* and *ranks* of X (cf. §1) are special cases, corresponding to partitions $\lambda = (1, \ldots, 1, 0, \ldots, 0)$ and their conjugates $\widetilde{\lambda} = (i, 0, \ldots, 0)$.

6.4. Symmetric and skew-symmetric loci

There are similar formulas for bundle maps $\sigma \colon E^\vee \longrightarrow E$ which are symmetric ($\sigma^\vee = \sigma$) or skew-symmetric ($\sigma^\vee = -\sigma$). Such correspond to sections t_σ of $S^2 E$ or $\bigwedge^2 E$. The locus $D_k(\sigma)$ is defined as in §6.1, but now its expected dimension m is, for $k \leq e = \text{rank}(E)$,
$$m = \dim(X) - \binom{e - k + 1}{2} \qquad \text{(symmetric)},$$
$$m = \dim(X) - \binom{e - k}{2} \qquad \text{(skew-symmetric, } k \text{ even)}.$$
There are classes denoted $\mathbf{D}_k^s(\sigma)$ or $\mathbf{D}_k^{ss}(\sigma)$ in $A_m(D_k(\sigma))$ in each of these cases. The analogous formulas are
$$\mathbf{D}_k^s(\sigma) = 2^d \Delta_{d, d-1, \ldots, 1}(c(E)) \cap [X],$$
$$\mathbf{D}_k^{ss}(\sigma) = \Delta_{d-1, \ldots, 1}(c(E)) \cap [X].$$
These formulas also date from Giambelli; modern versions have been given by Barth, Tjurin, Józefiak-Lascoux-Pragacz, Harris-Tu, and Damon. A particularly simple treatment has recently been given by Pragacz.[9] The calculations depend on calculating Gysin push-forwards for $\pi \colon G_d(E) \longrightarrow X$ as in §6.1. All such push-forwards are known "in theory", but it requires ingenuity to find useful general formulas. One such [35] is, for any $\lambda = (\lambda_1, \ldots, \lambda_d)$, $\nu = (\nu_1, \ldots, \nu_k)$, $\alpha \in A_* X$,
$$\pi_*\big(\Delta_\lambda(s(S)) \cdot \Delta_\nu(s(Q)) \cap \pi^* \alpha\big) = \Delta_\mu(s(E)) \cap \alpha,$$
where $\mu = (\lambda_1 - k, \lambda_2 - k, \ldots, \lambda_d - k, \nu_1, \ldots, \nu_k)$.

For applications, cf. Harris-Tu [29], one needs generalizations to symmetric or skew-symmetric bundle maps $\sigma\colon E^\vee \longrightarrow E \otimes L$, for L a line bundle on X. Lascoux and Pragacz also make these formulas explicit, as follows. Given partitions $\lambda = (\lambda_1, \ldots, \lambda_e)$ and $\mu = (\mu_1, \ldots, \mu_e)$, say that $\mu \leq \lambda$ if $\mu_i \leq \lambda_i$ for $1 \leq i \leq e$, and define
$$d_{\lambda\mu} = \det \left| \binom{\lambda_i + e - i}{\mu_j + e - j} \right|_{1 \leq i,j \leq e}.$$
Set $\varepsilon = (d, d-1, \ldots, 1)$ and $\delta = (d-1, d-2, \ldots, 1)$. Then for $d = e-k$, $e = \operatorname{rank}(E)$,
$$\mathbf{D}_k^s(\sigma) = 2^{-\binom{d}{2}} \sum_{\lambda \leq \varepsilon} 2^{|\lambda|} d_{\varepsilon\lambda} \Delta_{\widetilde{\lambda}}(c(E)) c_1(L)^{\binom{d+1}{2} - |\lambda|},$$
$$\mathbf{D}_k^{ss}(\sigma) = 2^{-\binom{d-1}{2}} \sum_{\lambda \leq \delta} 2^{|\lambda|} d_{\delta\lambda} \Delta_{\widetilde{\lambda}}(c(E)) c_1(L)^{\binom{d-1}{2} - |\lambda|}.$$

Here $\widetilde{\lambda}$ is the conjugate partition of λ. When $L = M^{\otimes 2}$, they follow from the preceding cases and the formal identity [39]
$$\Delta_{\widetilde{\lambda}}(c(E \otimes M)) = \sum_{\mu \leq \lambda} d_{\lambda\mu} \Delta_{\widetilde{\mu}}(c(E)) c_1(M)^{|\lambda| - |\mu|}.$$

The case for general L can be deduced from this case.[10]

CHAPTER 7

Refinements

7.1. Dynamic intersections

Consider our basic intersection theory setup

$$\begin{array}{ccc} W & \hookrightarrow & V \\ \downarrow & & \downarrow \\ X & \hookrightarrow & Y \\ & f & \end{array}$$

with f a regular embedding of codimension d, V an n-dimension variety, $W = X \cap V$. The normal cone $C = C_W V$ is embedded in the normal bundle $N = N_X Y$ to X in Y. Let

$$[C] = \sum m_i [C_i]$$

be the cycle of C. Each irreducible component C_i of C is a subcone of N; let $Z_i = s_N^{-1}(C_i)$ be the support of C_i. Let N_i be the restriction of N to Z_i. Then C_i is a subvariety of N_i, and we may set

$$\alpha_i = s_{N_i}^*[C_i] \in A_{n-d}(Z_i).$$

By construction, the class $\sum m_i \alpha_i$ represents the intersection product $X \bullet V$ in $A_{n-d}(W)$. Whenever some $Z_i \neq W$ there is more information in the classes α_i, with their multiplicities m_i, than in the class $\sum m_i \alpha_i$ on W. For any closed subset Z of W, set

$$(X \bullet V)^Z = \sum_{Z_i \subset Z} m_i \alpha_i \in A_{n-d}(Z),$$

and call $(X \bullet V)^Z$ the *part of $X \bullet V$ supported on Z*.

One way to refine this class further is to have a section s of the bundle N other than the zero section. Then $s^![C_i]$ is a well-defined class on $s^{-1}(C_i) \subset Z_i$, which *refines* α_i (i.e., $s^![C_i]$ maps to α_i by the inclusion of $s^{-1}(C_i)$ in Z_i). Suppose N is generated by a finite-dimensional space Γ of sections. One can show that for any closed $Z \subset W$ there is a nonempty open $\Gamma(Z) \subset \Gamma$ such that for all $s \in \Gamma(Z)$, $\dim s^{-1}(C) = n - d$, so $s^![C]$ is a well-defined $(n-d)$-cycle, and the part of $s^![C]$ contained in Z is precisely $(X \bullet V)^Z$.

Suppose the embedding $X \longrightarrow Y$ is deformed to a family $\mathcal{X} \longrightarrow Y \times T$ of embeddings; we assume T is a nonsingular curve, $\mathcal{X}$ is flat over T, the embedding of $\mathcal{X}$ in $Y \times T$ is regular, and the embedding $X_0 \longrightarrow Y \times \{0\}$ over $0 \in T$ is the given embedding. This deformation determines in a well-known way a *Kodaira-Spencer* section of the normal bundle N, which we denote by $s_{\mathcal{X}}$.

If the generic intersection of X_t with V is proper, one may define a *limit intersection cycle* $\lim_{t \to 0} X_t \bullet V$ as follows. Consider the fibre square:

$$\begin{array}{ccc} \mathcal{W} & \longrightarrow & V \times T \\ \downarrow & & \downarrow \\ \mathcal{X} & \longrightarrow & Y \times T \end{array}$$

The components of $\mathcal{W}$ that project dominantly to T have relative dimension $n - d$ over T. The intersection cycle $\mathcal{X} \bullet (V \times T)$ in $A_{n-d+1}(\mathcal{W})$ therefore specializes to a well-defined cycle on the fibre over 0 (cf. §3.4); this cycle is denoted $\lim_{t \to 0} X_t \bullet V$.

One can show that $\lim_{t \to 0} X_t \bullet V$ is supported on $s_{\mathcal{X}}^{-1}(C)$, and that this limit class refines $s^![C]$. It follows that, if Z is given, then for any deformation $\mathcal{X}$ for which $s_{\mathcal{X}}$ belongs to $\Gamma(Z)$,

$$\lim_{t \to 0} X_t \bullet V = s_{\mathcal{X}}^![C].$$

In particular, the part of the cycle $\lim_{t \to 0} X_t \bullet V$ supported on Z represents $(X \bullet V)^Z$, for sufficiently general deformations, cf. [40], [16].

For example, if $X = H_1 \times \cdots \times H_d$ and $Y = \mathbb{P}^n \times \cdots \times \mathbb{P}^n$ (d copies), with H_i hypersurfaces in $\mathbb{P}^n$, one may construct such deformations by varying equations for the H_i, as in §1. In case $d = n$, $V = \mathbb{P}^n$, it follows that the degree of $(X \bullet V)^Z$ is the number $j(Z)$ constructed by the Severi-Lazarsfeld method. Thus the (refined) *static* construction of intersection products (using normal cones) yields the same information as the *dynamic* construction (using deformations).[11]

7.2. Rationality of solutions

In much of our geometric discussion, we have been tacitly assuming that the ground field is the complex numbers, or at least algebraically closed. No such assumptions are needed for the basic constructions, however. If one begins with varieties and cycles defined over a given ground field K, all our operations can be carried out with cycles defined over K.

The *degree* of a zero-cycle $\sum n_i[P_i]$ on X is $\sum n_i[R(P_i):K]$, where $R(P)$ denotes the residue field of the local ring of X at P. For a zero-cycle or class α on a complete variety X over K, we let $\int \alpha$ denote its degree.

Suppose $V_1, \ldots, V_r$ are subvarieties of a complete smooth variety X over K, with $\sum \mathrm{codim}(V_i, X) = \dim X$. Then our construction produces a cycle class

$$V_1 \bullet \ldots \bullet V_r \in A_0\left(\bigcap V_i\right)$$

whose degree is $\int [V_1] \bullet \ldots \bullet [V_r]$.

For example, if $K = \mathbb{R}$, and $\int [V_1] \bullet \ldots \bullet [V_r]$ is *odd*, it follows that $\bigcap V_i$ must contain real points. Indeed, a zero-cycle $\sum n_i[P_i]$ on $\bigcap V_i$ which represents $V_1 \bullet \ldots \bullet V_r$ cannot have all $R(P_i) = \mathbb{C}$. Note that this argument can be used on each connected component of $\bigcap V_i$. For example, if certain points of proper intersections are known, their contributions can be subtracted; if an odd number remains, there are additional real points in $\bigcap V_i$. Similarly one may subtract contributions from components of excess intersection.

A pleasant application of these ideas is to a simple algebraic treatment of the Borsuk-Ulam problem [2]. Let S^n be the sphere $X_0^2 + \cdots + X_n^2 = 1$ in $\mathbb{R}^{n+1}$. If

$g_1, \ldots, g_n$ are odd polynomials in $\mathbb{R}[X_0, \ldots, X_n]$, then there is a point $x \in S^n$ such that all $g_i(x) = 0$. To prove this, for any odd g of degree d, set

$$g^* = \sum (x_0^2 + \cdots + x_n^2)^{(d-j)/2} g^{(j)},$$

where $g^{(j)}$ is the homogeneous part of g of (odd) degree j. By the previous paragraph, $g_1^*, \ldots, g_n^*$ have a common nontrivial solution $(x_0, \ldots, x_n)$. Multiplying by a positive scalar, one may assume this point is in S^n, in which case it is the required solution. It follows that for any n polynomials (or continuous functions, by approximation), there is a point $x \in S^n$ such that each takes the same value at antipodal points; one applies the preceding to the odd parts of the functions.

For $X = \mathbb{P}^n$ one may also prove such results by using deformations and the compactness of $\mathbb{P}^n(\mathbb{R})$. The approach with refined intersections is simpler; it works for any X, and gives analogous results for any field all of whose finite extensions have degrees that are powers of a fixed prime.

The question of how many solutions of real equations can be real is still very much open, particularly for enumerative problems. For example, how many of the 3264 conics tangent to five general (real) conics can be real?[12]

7.3. Residual intersections

In our basic situation for constructing intersection products (§7.1), there may be a distinguished subscheme D of the intersection scheme W. A natural candidate for the contribution of D to the intersection product $X \bullet V$ is the class

$$\{c(N_X Y) \cap s(D, V)\}_m \in A_m D,$$

$m = \dim(V) - d$, $d = \operatorname{codim}(X, Y)$. Our object is to construct a residual scheme R and a class, denoted $\mathbf{R}$, in $A_m(R)$ so that one has a

Residual intersection formula $\qquad X \bullet V = \{c(N_X Y) \cap s(D, V)\}_m + \mathbf{R}$

in $A_m W$.

We have a diagram

$$\begin{array}{ccccc} D & \xrightarrow{a} & W & \xrightarrow{j} & V \\ & & \downarrow g & & \downarrow f \\ & & X & \xrightarrow{i} & Y \end{array}$$

with i a regular embedding, $W = f^{-1}(X)$. Assume first that the composite ja embeds D as a Cartier divisor on V. The *residual scheme* R to D in W is the subscheme of V whose local equations are obtained by dividing local equations for W in V by a local equation for D in V; then $W = D \cup R$, with ideal sheaves on V related by

$$\mathfrak{I}(W) = \mathfrak{I}(D) \cdot \mathfrak{I}(R).$$

Set $E = g^* N_X Y \otimes j^* \mathcal{O}_V(-D) = g^* N_X Y \otimes (N_D V)^\vee$. One verifies that the normal cone $C_R V$ is a subcone of the restriction E_R of E to R. Then one may define the *residual class* $\mathbf{R}$ to be the intersection class of the n-cycle $[C_R V]$ by the zero section of the bundle E_R:

$$\mathbf{R} = s^*_{E_R}[C_R V] = \{c(E) \cap s(R, V)\}_m.$$

With these definitions, the residual intersection formula is valid. To prove it, one blows up V along Z to reduce to the case where $W = D + R$ as a divisor, in which case the excess intersection formula applies; see [**16**, §9] for details.

For example, if the embedding of R in V is regular of codimension d', then

$$\mathbf{R} = c_{d-d'}(E/N_R V) \cap [R].$$

In particular, if $d' = d$, then $\mathbf{R} = [R]$.

For arbitrary D, one can blow up V along D to achieve the situation just studied. Let $\pi \colon \widetilde{V} \longrightarrow V$ be this blow-up, $\pi^* D = \widetilde{D}$ and $\pi^{-1}(W) = \widetilde{W}$. Let $\widetilde{R}$ be the residual scheme to $\widetilde{D}$ in $\widetilde{W}$ and $\widetilde{\mathbf{R}}$ the residual class in $A_m \widetilde{R}$ just constructed. If one sets $R = \pi(\widetilde{R})$ and $\mathbf{R} = \pi_*(\widetilde{\mathbf{R}})$, the desired residual intersection formula results.

7.4. Multiple point formulas

Laksov developed a version of this residual intersection formula to prove a double point formula for a morphism $f \colon X \longrightarrow Y$ of nonsingular varieties of dimensions n and m. In this case $(f \times f)^{-1}(\Delta_Y)$ contains Δ_X; the residual scheme will be the locus $D'(f)$ of double point pairs. If X is complete either projection from $X \times X$ to X maps $D'(f)$ onto the *double point* locus $D(f)$ in X. The projection of the residual intersection class $\mathbf{R}$ is the *double point class*, denoted $\mathbf{D}(f)$, in $A_{2n-m}(D(f))$. One deduces from the residual intersection formula the

Double point formula $\qquad \mathbf{D}(f) = f^* f_*[X] - \bigl(c(f^* T_Y) c(T_X)\bigr)_{m-n} \cap [X].$

For example, if X is a curve of genus g, $Y = \mathbb{P}^2$, and f maps X birationally onto a curve of degree n, one has the classical formula

$$\deg \mathbf{D}(f) = (n-1)(n-2) - 2g.$$

By the preceding section, $\mathbf{D}(f)$ is constructed from the residual scheme $\widetilde{R}$ to the exceptional divisor $\mathbb{P}(T_X)$ in the blow-up of $X \times X$ along the diagonal Δ_X. When this scheme $\widetilde{R}$ has the expected dimension $2n - m$, then $\mathbf{D}(f)$ is the projection of the cycle $[\widetilde{R}]$. Note that $\widetilde{R}$ may have components inside $\mathbb{P}(T_X)$, as happens e.g. for plane curves with cusps. When $m = n + 1$ and f maps X birationally and finitely onto its image in Y, one expects $\mathbf{D}(f)$ to be the cycle determined by the *conductor* ideal, but this has only been proved for $n = 1$.[13]

For triple point and higher multiple point formulas the situation is more complicated; however, when f is a proper immersion which is *completely regular* (i.e., for any distinct points $x_i \in X$ with the same image $y \in Y$, the images of the tangent spaces $T_{x_i} X$ are in general position in $T_y Y$), the answer is quite simple. Let Y_k be the set of points in Y which are the images of k or more distinct points of X, and let $X_k = f^{-1}(Y_k)$, both with reduced scheme structure. Then one has an inductive formula of Herbert:

$$[X_k] = f^*[Y_{k-1}] - c_d(f^* T_Y / T_X) \cap [X_{k-1}],$$

$d = \dim Y - \dim X$. Indeed, following Ronga, if X_k^* denotes the set of unordered k-tuples of distinct points of X with the same image on Y, one has a fibre square

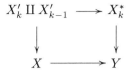

with X_j' mapping birationally onto X_j. An application of the excess intersection formula (§4.3) then yields Herbert's formula.

For more general mappings significant progress has been made, primarily by Kleiman, Le Barz, and Ran. Their results are most satisfactory when the multiple points occur only in a "curvilinear" way; they can be used to deduce enumerative formulas for secant lines to varieties in projective space.[14]

The excess intersection formulas can also be used to study fixed points of correspondences on a nonsingular complete variety X. If Γ is a variety, cycle, or equivalence class of cycles on X, with $\dim \Gamma = \dim X$, then the *virtual* number of fixed points is the intersection number $\int \Gamma \bullet \Delta$ of Γ with the diagonal Δ in $X \times X$. When $X = \mathbb{P}^n$, one recovers formulas of Pieri [49]. Conversely, Pieri's work may be seen as an important precursor of modern intersection theory.

CHAPTER 8

Positivity

8.1. Positivity of intersection products

When varieties meet properly their intersection product will be an *effective* cycle, i.e. a sum $\sum n_i[V_i]$ with $n_i \geq 0$. If two cycles are equivalent to effective cycles that meet properly, their product is represented by (equivalent to) a positive (or zero) cycle. For general excess intersections, however, this is not possible: if E is the exceptional divisor of the blow-up of a nonsingular surface at a point, then $\int E \bullet E = -1$.

From our construction of intersection products via cones in normal bundles, it is natural to expect that suitable positivity of the normal bundle will guarantee the positivity of intersection products.

Recall that a line bundle L on a variety X is *ample* if some positive power $L^{\otimes n}$ is the pull-back of $\mathcal{O}(1)$ for a projective embedding of X in projective space. If α is a k-cycle on X, define

$$\deg_L(\alpha) = \int_X c_1(L)^k \cap \alpha.$$

For a subvariety V of X, let $\deg_L(V) = \deg_L[V]$. Since hyperplanes can always be moved to meet subvarieties properly, $\deg_L(\alpha) > 0$ whenever α is equivalent to a *positive* cycle, i.e. nonzero effective cycle.

A vector bundle E on X is *ample* if the canonical line bundle $\mathcal{O}_{E^\vee}(1)$ on $\mathbb{P}(E^\vee)$ is an ample line bundle; note that $\mathcal{O}_{E^\vee}(1)$ is a quotient bundle of the pull-back of E to $\mathbb{P}(E^\vee)$. In general, ampleness is preserved by direct sums; by tensor, exterior, and symmetric products; by passing to quotient bundles; and by pull-backs by finite morphisms.

To investigate the positivity of intersection products it suffices to consider the intersection class of an irreducible cone C with the zero section in a vector bundle E on a variety X. Let $\alpha = s_E^*[C]$ be this intersection class. We assume that $\dim(C) \geq \operatorname{rank}(E)$. Fix an ample line bundle L on X.

THEOREM. *(a)* If E is generated by its sections, then α is represented by an effective cycle.

(b) If $E \otimes L^\vee$ is generated by its sections, and $\operatorname{Supp}(C) = X$, then $\deg_L(\alpha) \geq \deg_L(X)$.

(c) If E is ample and generated by its sections, then α is represented by a positive cycle.

(d) If E is ample, then $\deg_L(\alpha) > 0$.

Of these statements, (a) and (b) are quite easy to prove; they follow from the case where E is trivial, and one may induct on the rank. We refer to [**16**, §12] for the proofs of (a)–(c).

The most difficult is (d). For example, when C is the zero section, and $\dim X = \mathrm{rank}(E) = n$, then $\alpha = c_n(E) \cap [X]$. The assertion that $c_n(E) > 0$ is a theorem of Bloch and Gieseker [7]. Their proof, valid in characteristic zero, used resolution of singularities and the hard Lefschetz theorem. By using intersection homology [27] one may avoid resolution of singularities and extend the result to arbitrary characteristic, but we do not know a more elementary proof of the key assertion that $c_n(E) \neq 0$. We do not know, in (d), if some positive multiple of α can be represented by a positive cycle. For the proof of (d), see [20].

The theorem applies to the intersection products $X \bullet V \in A_m(W)$ constructed from our basic construction, provided the pull-back N of $N_X Y$ to W has the required positivity. For example, if $N_X Y$ is ample, and V is a subvariety of Y with $\dim V \geq \mathrm{codim}(X, Y)$, such that $[V]$ is equivalent to a positive cycle whose support meets X, then V must itself meet X. Indeed, $\deg_L(X \bullet V) = \deg_L(X \bullet \alpha) > 0$.

If X is nonsingular and its tangent bundle T_X is generated by its sections, it follows from (a) that all intersections of effective cycles have effective representatives. Indeed, the normal bundle to the diagonal embedding of X in $X \times \cdots \times X$ (r copies) is the direct sum $T_X \oplus \cdots \oplus T_X$ ($r - 1$ copies).

If $X = \mathbb{P}^n$ and $L = \mathcal{O}(1)$, then $T_X \otimes L^\vee$ is generated by its sections, so (b) holds for all intersections on $\mathbb{P}^n$. Let $V_1, \ldots, V_r$ be subvarieties of $\mathbb{P}^n$, and let $V_1 \bullet \ldots \bullet V_r$ be the intersection product constructed from the diagram

$$\begin{array}{ccc} \bigcap V_j & \longrightarrow & V_1 \times \cdots \times V_r \\ \downarrow & & \downarrow \\ X & \xrightarrow{\delta} & X \times \cdots \times X \end{array}$$

by the prescription of §7.1. If $[C] = \sum m_i [C_i]$ is the cycle of the normal cone to $\bigcap V_j$ in $V_1 \times \cdots \times V_r$, then we have a decomposition

$$V_1 \bullet \ldots \bullet V_r = \sum m_i \alpha_i$$

with α_i a class on $Z_i = \mathrm{Supp}(C_i)$. By the theorem, each α_i is represented by a positive cycle, and $\deg(\alpha_i) \geq \deg(Z_i) > 0$. In particular,

$$\prod \deg(V_i) = \sum m_i \deg(\alpha_i) \geq \sum m_i \deg(Z_i).$$

Note that each irreducible component of $\bigcap V_j$ appears as some Z_i, and each $m_i \geq 1$, so this refines the Bézout theorem of §2.3.

8.2. Positive polynomials and degeneracy loci

Let $\sigma \colon E \longrightarrow F$ be a homomorphism of vector bundles of ranks e and f on an n-dimensional variety X. For $k \leq \min(e, f)$, the expected dimension of the degeneracy locus $D_k(\sigma)$ was seen in §6.1 to be $m = n - (e - k)(f - k)$.

PROPOSITION. *Assume $E^\vee \otimes F$ is ample, and $m \geq 0$. Then $D_k(\sigma) \neq \emptyset$, and for any ample line bundle L on X,*

$$\deg_L \left(\Delta^{(e-k)}_{f-k} (c(F - E)) \right) > 0.$$

To prove this, let $H = \mathrm{Hom}(E, F) = E^\vee \otimes F$, and let $D_k \subset H$ be the cone of maps of rank $\leq k$ (§6.1). Then σ corresponds to a section t_σ on H, and the degeneracy class $\mathbf{D}_k(\sigma)$ is the intersection class $t_\sigma^! [D_k]$. Since the normal bundle to

8.2. POSITIVE POLYNOMIALS AND DEGENERACY LOCI

t_σ is H, which is assumed to be ample, the theorem of §8.1 yields $\deg_L(\mathbf{D}_k(\sigma)) > 0$. The Giambelli-Thom-Porteous formula for $\mathbf{D}_k(\sigma)$ completes the proof.

A similar construction shows that if E is any ample vector bundle of rank e on an n-dimensional variety X, then for any partition λ of n with $e \geq \lambda_1 \geq \lambda_2 \geq \cdots \geq 0$,
$$\int \Delta_\lambda(c(E)) > 0.$$

To prove this one takes a vector space V of dimension $n+e$, and a flag of subspaces $V_1 \subset V_2 \subset \cdots \subset V$ with $\dim(V_i) = e + i - \lambda_i$. Let $H = \text{Hom}(V_X, E)$, and set
$$\Omega_\lambda = \{\phi \in H \mid \dim \text{Ker}(\phi) \cap V_i \geq i, \ i = 1, \dots\}.$$

Then H is ample, and σ corresponds to a section t_σ of H. By the determinantal formula, $\Delta_\lambda(c(E)) \cap [X] = t_\sigma^*[\Omega_\lambda]$, whose degree is positive by the theorem of §8.1.

Any polynomial $P(c_1, \dots, c_e) \in \mathbb{Q}[c_1, \dots, c_e]$ of weight n can be written uniquely in the form
$$P = \sum a_\lambda \Delta_\lambda(c);$$
the sum over partitions λ with $e \geq \lambda_1 \geq \cdots \geq \lambda_n \geq 0$, and $\sum \lambda_i = n$.

THEOREM ([20]). *If $a_\lambda \geq 0$ for all λ, and some $a_\lambda > 0$, then*
$$\int_X P(c_1(E), \dots, c_k(E)) > 0$$
for all ample vector bundles E of rank e on all n-dimensional varieties X. Conversely, if some $a_\lambda < 0$, there is an ample E on some X for which $\int_X P(c(E)) < 0$.

To see the last statement, let V be a Schubert variety representing the *dual* class to $\{\lambda_1, \dots, \lambda_n\}$ in $G = G_n(\mathbb{P}^{n+e})$. Then
$$\int_V P(c(Q)) = a_\lambda < 0$$
for Q the universal quotient bundle on G. Let L be a very ample line bundle on G. For any $k > 0$ there is a finite surjective morphism $f: X \longrightarrow V$ such that $f^*L = M^{\otimes k}$ for a line bundle M on X. Then $E = f^*Q \otimes M$ is ample on X, and if $F(t)$ is the polynomial
$$F(t) = \int_V P(c(Q \otimes L^{\otimes t})),$$
then $\int_X P(c(E)) = \deg(X/V) F(1/k)$; this is negative for sufficiently large k, since $F(0) = a_\lambda < 0$.[15]

A similar analysis shows the positivity of products of Schur polynomials in Chern classes in two or more ample bundles, and the existence of degeneracy loci $D_k^s(\sigma)$ and $D_k^{ss}(\sigma)$ for symmetric or skew-symmetric bundle maps $\sigma: E^\vee \longrightarrow E \otimes L$, when the expected dimension is nonnegative, and $S^2(E) \otimes L$ or $\bigwedge^2 E \otimes L$ is ample.

Although we have been using rational equivalence, the natural equivalence for most of these questions is numerical equivalence. Two cycles α, α' on a complete variety X may be said to be *numerically equivalent* if $\int P \cap \alpha = \int P \cap \alpha'$ for all polynomials P in Chern classes of vector bundles on X. When X is nonsingular, it follows from Riemann-Roch that this is equivalent to requiring $\int \beta \bullet \alpha = \int \beta \bullet \alpha'$ for all cycles β on X.

If the expected dimension m of $D_k(\sigma)$ is at least 1, and $E^\vee \otimes F$ is ample, then $D_k(\sigma)$ must be *connected* [19]. The analogous assertion is open for symmetric and skew-symmetric bundle maps. These assertions would follow from the general conjecture that for any k-dimensional subvariety V of an ample bundle of rank e, and any section s of V, $s^{-1}(V)$ is connected, provided $k > e$. Note that the nonemptiness of $s^{-1}(V)$ follows from the theorem of §8.1, for any $k \geq e$.

8.3. Intersection multiplicities

Let $V_1, \ldots, V_r$ be subvarieties of an n-dimensional nonsingular variety X meeting properly at a point P (assumed to be rational over the ground field). Let $i(P) = i(P, V_1 \bullet \ldots \bullet V_r; X)$ be the intersection multiplicity. By shrinking X, we may assume the V_i intersect only at P.

Let $\pi \colon \widetilde{X} \longrightarrow X$ be the blow-up of X at P, $E = \mathbb{P}^{n-1}$ the exceptional divisor. Let $\widetilde{V}_i \subset \widetilde{X}$ be the blow-up of V_i at P, i.e., the proper transform. Note that $\widetilde{V}_i \cap E$ is the projective tangent cone $\mathbb{P}(C_P V_i)$, whose degree in $E = \mathbb{P}^{n-1}$ is the multiplicity $e_P V_i$ of V_i at P (§4.2). Note also that the intersection product $\widetilde{V}_1 \bullet \ldots \bullet \widetilde{V}_r$ on $\widetilde{X}$ is a well-defined class in $A_0(E)$. Then

$$(*) \qquad i(P) = e_P(V_1) \bullet \ldots \bullet e_P(V_r) + \int_E \widetilde{V}_1 \bullet \ldots \bullet \widetilde{V}_r.$$

For two curves on a surface the curves $\widetilde{V}_1$ and $\widetilde{V}_2$ must intersect properly, and one may continue blowing up; this results in Noether's formula for $i(P)$ as the sum of the products of the multiplicities at all infinitely near points. (One can prove $(*)$ in general by using either deformation to the normal bundle to P in X, or the residual intersection formula.)

In general, the $\widetilde{V}_i$ need not intersect properly. Since intersections on $\widetilde{X}$ can be negative, it is not so obvious that $\widetilde{V}_1 \bullet \ldots \bullet \widetilde{V}_r$ must be nonnegative. Using the theorem of §8.1, however, one can show that there is a decomposition

$$\widetilde{V}_1 \bullet \ldots \bullet \widetilde{V}_r = \sum m_i \alpha_i$$

with $m_i > 0$, α_i a cycle on a subvariety Z_i of $\bigcap \widetilde{V}_i = \bigcap \mathbb{P}(C_P V_i)$,

$$\deg(\alpha_i) \geq \deg Z_i > 0.$$

The union of the Z_i is $\bigcap \widetilde{V}_i$. In particular, the "error term" $\int \widetilde{V}_1 \bullet \ldots \bullet \widetilde{V}_r$ is bounded below by the sum of the degrees of the irreducible components of $\bigcap \mathbb{P}(C_P V_i) \subset E = \mathbb{P}^{n-1}$.

When the V_i are hypersurfaces the proof is an easy application of the theorem, since the normal bundle to $\widetilde{V}_i$ in $\widetilde{X}$ is ample on $\widetilde{V}_i \cap E$. For general V_i the proof is more complicated, since the normal bundle to the diagonal embedding of X in $\widetilde{X} \times \cdots \times \widetilde{X}$ is *not* ample near E. See [**16**, §12.4] for details.

CHAPTER 9

Riemann-Roch

9.1. The Grothendieck-Riemann-Roch theorem

If E is a vector bundle on a complete variety X, let $\chi(E)$ denote its Euler characteristic:
$$\chi(E) = \sum (-1)^i \dim H^i(X, E).$$
Motivated by some ingenious calculations of Todd, Hirzebruch discovered the *Hirzebruch-Riemann-Roch* formula (HRR) for expressing $\chi(E)$ in terms of Chern classes of E and of the tangent bundle of a nonsingular variety X:
$$\chi(E) = \int_X \operatorname{ch}(E) \cdot \operatorname{td}(T_X).$$
Here ch and td denote the Chern character and Todd class respectively. The Chern character $\operatorname{ch}(E)$ of a vector bundle E of rank e on a variety X is the sum
$$\operatorname{ch}(E) = e + c_1 + \tfrac{1}{2}(c_1^2 - 2c_2) + \tfrac{1}{6}(c_1^3 - 3c_1c_2 + 3c_3) + \cdots$$
$$= \sum_{k \geq 0} p_k/k!,$$
where $c_i = c_i(E)$, and p_k is the sum $x_1^k + \cdots + x_e^k$ in Chern roots x_i of E; explicitly,

$$p_k = \det \begin{vmatrix} c_1 & 1 & 0 & \cdots & 0 \\ 2c_2 & c_1 & 1 & \cdots & 0 \\ \vdots & & & & \vdots \\ & & & c_1 & 1 \\ kc_k & c_{k-1} & \cdots\cdots & & c_1 \end{vmatrix}.$$

Then $\operatorname{ch}(E) = \operatorname{ch}(E') + \operatorname{ch}(E'')$ if $0 \to E' \to E \to E'' \to 0$ is exact, and $\operatorname{ch}(E \otimes F) = \operatorname{ch}(E) \cdot \operatorname{ch}(F)$. Similarly the *Todd class* $\operatorname{td}(E)$ is defined by
$$\operatorname{td}(E) = 1 + \tfrac{1}{2}c_1 + \tfrac{1}{12}(c_1^2 + c_2) + \tfrac{1}{24}c_1 c_2 + \cdots$$
$$= \prod_{i=1}^{e} x_i/(1 - \exp(-x_i)).$$
They are related by the formal identity

(i) $$\sum_{i=0}^{e} (-1)^i \operatorname{ch}(\wedge^i E^\vee) = c_e(E) \cdot \operatorname{td}(E)^{-1}.$$

Note that $\operatorname{td}(E) = \operatorname{td}(E') \cdot \operatorname{td}(E'')$ if $0 \to E' \to E \to E'' \to 0$ is exact.

Both χ and ch are additive on exact sequences of vector bundles, the former by the long exact cohomology sequence. Let $K^\circ X$ denote the *Grothendieck group* of

(algebraic) vector bundles on X; it is the free abelian group on isomorphism classes $[E]$ of vector bundles on X, modulo relations
$$[E] = [E'] + [E'']$$
for any exact sequence $0 \to E' \to E \to E'' \to 0$ on X. If X is complete, then χ determines a homomorphism from $K^\circ X$ to $\mathbb{Z}$. For a nonsingular complete variety X, set $K(X) = K^\circ X$, $A(X) = A^*X$, and let
$$\tau: K(X) \longrightarrow A(X)_\mathbb{Q} = A(X) \otimes \mathbb{Q}$$
be the homomorphism given by $\tau(E) = \mathrm{ch}(E) \cdot \mathrm{td}(T_X)$. So HRR reads: $\chi(E) = \int \tau(E)$.

If $f: X \longrightarrow Y$ is a closed embedding of nonsingular varieties, there is an induced homomorphism f_* from $K(X)$ to $K(Y)$, determined by
$$f_*[E] = \sum_{i=0}^{m}(-1)^i [F_i]$$
if $0 \to F_m \to F_{m-1} \to \cdots \to F_0 \to f_*E \to 0$ is a resolution of the sheaf f_*E by vector bundles F_i. Assume X and Y are complete, and consider the diagram:

$$\begin{array}{ccccc}
K(X) & \xrightarrow{f_*} & K(Y) & \xrightarrow{\chi} & \mathbb{Z} \\
{\scriptstyle \tau}\downarrow & & {\scriptstyle \tau}\downarrow & & \downarrow \\
A(X)_\mathbb{Q} & \xrightarrow{f_*} & A(Y)_\mathbb{Q} & \longrightarrow & \mathbb{Q}
\end{array}$$

One sees that to prove HRR on X it suffices to know HRR on Y (so the right square commutes), and the commutativity of the left square. When X is projective, one may take $Y = \mathbb{P}^n$. Then $K(\mathbb{P}^n)$ is generated by $[\mathcal{O}(i)]$ for $i = 0, \ldots, n$, and the verification of HRR for these line bundles amounts to a formal identity. So the essential part of the proof of HRR, for X projective and nonsingular, is to verify the commutativity of the left square of the diagram. That is, for all $\alpha \in K(X)$,

(*) $\qquad \mathrm{ch}(f_*\alpha) \cdot \mathrm{td}(T_Y) = f_*\bigl(\mathrm{ch}(\alpha) \cdot \mathrm{td}(T_X)\bigr).$

Let N be the normal bundle to X in Y. Since td takes sums to products,
$$\mathrm{td}(N)^{-1} \cdot \mathrm{td}(f^*T_Y) = \mathrm{td}(T_X).$$
By the projection formula, (*) is equivalent to

(**) $\qquad \mathrm{ch}(f_*\alpha) = f_*\bigl(\mathrm{td}(N)^{-1} \cdot \mathrm{ch}(\alpha)\bigr).$

Let us first verify (**) on a simple example, where everything can be calculated explicitly: X is an arbitrary nonsingular variety, $Y = \mathbb{P}(N \oplus \mathbb{1})$, where N is an arbitrary vector bundle on X, and $f: X \longrightarrow Y$ is the zero-section embedding of X in N, followed by the open embedding of N in $\mathbb{P}(N \oplus \mathbb{1})$. Let $p: Y \longrightarrow X$ be the bundle projection, let Q be the universal quotient bundle on $\mathbb{P}(N \oplus \mathbb{1})$, and $d = \mathrm{rank}(N) = \mathrm{rank}(Q)$. Let s be the section of Q determined by the projection of the trivial factor in $p^*(N \oplus \mathbb{1})$ onto Q. The zero scheme of s is precisely X. It follows that for any $\beta \in A(Y)$,

(ii) $\qquad f_*(f^*\beta) = \beta \cdot f_*[X] = c_d(Q) \cdot \beta.$

The section s determines a *Koszul complex*
$$0 \longrightarrow \wedge^d Q^\vee \longrightarrow \wedge^{d-1} Q^\vee \longrightarrow \cdots \longrightarrow \wedge^1 Q^\vee \longrightarrow \mathcal{O}_Y \longrightarrow f_*\mathcal{O}_X \longrightarrow 0$$

9.1. THE GROTHENDIECK-RIEMANN-ROCH THEOREM

which is a resolution of $f_*\mathcal{O}_X$. It follows that for any locally free sheaf E on X, f_*E has a resolution
$$0 \longrightarrow \bigwedge^d Q^\vee \otimes p^*E \longrightarrow \cdots \longrightarrow \bigwedge^1 Q^\vee \otimes p^*E \longrightarrow p^*E \longrightarrow f_*E \longrightarrow 0.$$
Therefore,
$$\mathrm{ch}(f_*[E]) = \sum (-1)^i \mathrm{ch}(\bigwedge^i Q^\vee) \cdot \mathrm{ch}(p^*E).$$
From (i), the right side is $c_d(Q) \cdot \mathrm{td}(Q)^{-1} \cdot \mathrm{ch}(p^*E)$. Using (ii), and noting that $f^*Q = N$ and $f^*p^*E = E$, one has

(iii) $$\mathrm{ch}(f_*E) = f_*\big(\mathrm{td}(N)^{-1} \cdot \mathrm{ch}(E)\big)$$

which proves (**) in this case. This model also shows, via identity (i), where the Todd classes come from.

To prove (**) in general, consider the deformation to the normal bundle (§2.6). In order to have a projective parameter space, we deform from the given embedding at $0 \in \mathbb{P}^1$ to the normal bundle embedding at $\infty \in \mathbb{P}^1$; i.e., M is the blow-up of $Y \times \mathbb{P}^1$ along $X \times \{\infty\}$:

$$\begin{array}{ccccccc}
X & \xrightarrow{\overline{f}} & \mathbb{P}(N \oplus 1) & + & \widetilde{Y} & = & M_\infty \longrightarrow \{\infty\} \\
{\scriptstyle i_\infty}\downarrow & & {\scriptstyle k}\searrow & {\scriptstyle \ell}\downarrow & {\scriptstyle j_\infty}\swarrow & & \downarrow \\
X \times \mathbb{P}^1 & & \xrightarrow{F} & M & & \xrightarrow{\rho} & \mathbb{P}^1 \\
{\scriptstyle i_0}\uparrow & & & {\scriptstyle j_0}\uparrow & & & \uparrow \\
X & & \xrightarrow{f} & Y & = & M_0 & \longrightarrow \{0\}
\end{array}$$

Here $\overline{f} \colon X \longrightarrow \mathbb{P}(N \oplus 1)$ is the preceding model. Now let E be any vector bundle on X. We must show that equation (**) holds. Let $q \colon M \longrightarrow Y$ be the composite of the blow-down map from M to $Y \times \mathbb{P}^1$ and the projection to Y. Since $qj_0 = \mathrm{id}_Y$, it will suffice to compute the image of $\mathrm{ch}(f_*E)$ in $A(M)_\mathbb{Q}$.

Let $\widetilde{E} = (\mathrm{pr})^*E$ be the pull-back of E to $X \times \mathbb{P}^1$, and let
$$0 \longrightarrow G_n \longrightarrow G_{n-1} \longrightarrow \cdots \longrightarrow G_1 \longrightarrow G_0 \longrightarrow F_*(\widetilde{E}) \longrightarrow 0$$
be a resolution of $F_*(\widetilde{E})$ on M. Since M is flat over $\mathbb{P}^1$, it follows that $j_0^*G_\bullet$ is a resolution of f_*E on Y and $j_\infty^*G_\bullet$ resolves $\overline{f}_*E$ on M_∞. In particular, since $\widetilde{Y}$ is disjoint from $\overline{f}(X)$, $\ell^*G_\bullet$ is acyclic.

Write $\mathrm{ch}(F_\bullet)$ in place of $\sum(-1)^i \mathrm{ch}(F_i)$, for any complex $F_\bullet$ of vector bundles. Using the projection formula, we have
$$j_{0*}(\mathrm{ch}(f_*E)) = j_{0*}(\mathrm{ch}(j_0^*G_\bullet)) = \mathrm{ch}(G_\bullet) \cdot j_{0*}[X].$$
Note that $j_{0*}[X] = k_*[\mathbb{P}(N\oplus 1)] + \ell_*[\widetilde{Y}]$, since $0 = [\mathrm{div}(\rho)] = [M_0] - [M_\infty]$. Therefore
$$\mathrm{ch}(G_\bullet) \cdot j_{0*}[X] = k_*(\mathrm{ch}(k^*G_\bullet)) + \ell_*(\mathrm{ch}(\ell^*G_\bullet))$$
$$= k_*(\mathrm{ch}(\widetilde{f}_*E)) + 0.$$
But $\mathrm{ch}(\overline{f}_*E)$ was calculated for the model. So we have
$$j_{0*}(\mathrm{ch}(f_*E)) = k_*\Big(\overline{f}_*\big(\mathrm{td}(N)^{-1} \cdot \mathrm{ch}(E)\big)\Big)$$

in $A(M)_{\mathbb{Q}}$. Applying q_* to both sides yields the required formula (**), since $qk\overline{f} = f$.

On an arbitrary variety X, let $K_\circ X$ denote the Grothendieck group of coherent sheaves on X. Tensor product makes $K^\circ X$ into a commutative ring, and $K_\circ X$ into a module over $K^\circ X$. If $f: Y \longrightarrow X$ is a morphism, the pull-back of vector bundles defines a ring homomorphism $f^*: K^\circ X \longrightarrow K^\circ Y$. If f is proper, there is a push-forward

$$f_*: K_\circ Y \longrightarrow K_\circ X, \quad f_*[\mathcal{F}] = \sum (-1)^i [R^i f_* \mathcal{F}].$$

Here $R^i f_* \mathcal{F}$ are Grothendieck's *higher direct images*: $R^i f_* \mathcal{F}$ is the sheaf associated to the presheaf $U \longrightarrow H^i(f^{-1}(U), \mathcal{F})$. One has the *projection formula* $f_*(f^*\alpha \otimes \beta) = \alpha \otimes f_*\beta$ for $\alpha \in K^\circ X$, $\beta \in K_\circ Y$.

For any X there is a homomorphism from $K^\circ X$ to $K_\circ X$ taking a bundle E to its sheaf of sections, or $\alpha \longmapsto \alpha \otimes [\mathcal{O}_X]$. If X is nonsingular, it follows from the fact that every coherent sheaf $\mathcal{F}$ on X has a finite resolution by locally free sheaves that this homomorphism is an isomorphism. Set $K(X) = K^\circ X \cong K_\circ X$, and $A(X) = A^* X = A_* X$; both K and A then become covariant for proper morphisms, as well as contravariant. Grothendieck realized that the Riemann-Roch problem could be formulated as the comparison of these two push-forwards, via the Chern character.

THEOREM (GRR [9]). *For any proper morphism $f: X \longrightarrow Y$ of nonsingular varieties, and any $\alpha \in K(X)$,*

$$\mathrm{ch}(f_*(\alpha)) \cdot \mathrm{td}(T_Y) = f_*\big(\mathrm{ch}(\alpha) \cdot \mathrm{td}(T_X)\big).$$

In other words the homomorphism τ is covariant for arbitrary proper morphisms. To prove that τ commutes with f_*, it suffices to factor f into a composite gh, such that the commutativity with g_* and h_* is known. For X quasiprojective, one can find such a factorization

$$X \xrightarrow{h} Y \times \mathbb{P}^n \xrightarrow{g} Y$$

with h a closed embedding and g the projection. We have proved GRR for closed embeddings; since $K(Y \times \mathbb{P}^n)$ is generated over $K(Y)$ by the classes $[\mathcal{O}(i)]$, the calculations that proved HRR for $\mathbb{P}^n$ also prove GRR for g. For extensions to varieties which may not be quasiprojective, see [17].

For a closed embedding $f: X \longrightarrow Y$, the same reasoning yields a Riemann-Roch formula without denominators, i.e. a formula

$$c_i(f_* E) = f_*\big(P_{i-d}(c(E), c(N))\big)$$

for certain polynomials P_j of weight j in the Chern classes of E and N, $d = \mathrm{codim}(X, Y)$. In particular, $c_i(f_*[\mathcal{O}_X]) = 0$ for $0 < i < d$, and

$$c_d(f_*[\mathcal{O}_X]) = (-1)^{d-1}(d-1)![X].$$

From this, or from a similar deformation argument, one proves a formula for the Chern classes $c_i(T_{\widetilde{Y}})$ of the blow-up $\widetilde{Y}$ of Y along X.

9.2. The singular case

The transformation τ from K to $A_{\mathbb{Q}}$ defined for nonsingular varieties in the previous section extends uniquely to the category of arbitrary varieties, in the

following sense. For every algebraic scheme X over a given field K there is a homomorphism
$$\tau_X \colon K_\circ(X) \longrightarrow A_*(X)_\mathbb{Q}$$
satisfying the properties:

(1) If $f \colon X \longrightarrow Y$ is proper and $\alpha \in K_\circ X$, then
$$f_* \tau_X(\alpha) = \tau_Y f_*(\alpha).$$

(2) If $\alpha \in K_\circ X$ and $\beta \in K^\circ X$, then
$$\tau_X(\beta \otimes \alpha) = \operatorname{ch}(\beta) \cap \tau_X(\alpha).$$

(3) If V is a subvariety of X, then
$$\tau_X(\mathcal{O}_V) = [V] + \text{lower terms}.$$

(4) If $f \colon X \longrightarrow Y$ is smooth with relative tangent bundle T_f, then for $\alpha \in K_\circ Y$,
$$\tau_X f^* \alpha = \operatorname{td}(T_f) \cdot f^* \tau_Y(\alpha).$$

These properties uniquely determine τ. Indeed, one only needs (1), (2), (3) for $V = \mathbb{P}^n$, and (4) for open embeddings, to characterize τ; note that none of these conditions refer to Todd classes. When X is nonsingular, $\tau_X(\alpha)$ is given by the formula $\operatorname{ch}(\alpha) \cdot \operatorname{td}(T_X)$ of §9.1.

For arbitrary X, and $\alpha \in K_\circ X$, one may often calculate $\tau_X(\alpha)$ by finding a proper morphism $\pi \colon X' \longrightarrow X$ with X' nonsingular and $\alpha' \in K_\circ X'$ with $\pi_* \alpha' = \alpha$. Then by (1),
$$\tau_X(\alpha) = \pi_* \bigl(\operatorname{ch}(\alpha') \cdot \operatorname{td}(T_{X'}) \bigr).$$
At least in characteristic zero such X', α' always exist, although X' may not be connected. Similarly using Chow's lemma, one may determine τ_X for arbitrary X from the construction of $\tau_{X'}$ for X' quasiprojective. What must be proved is that such constructions are independent of choices.

If X is a variety which admits a closed embedding in a nonsingular variety M, then τ_X may be constructed as follows. For a coherent sheaf $\mathcal{F}$ on X, let $E_\bullet$ be a resolution of $\mathcal{F}$ on M. Note that $\operatorname{ch}(E_\bullet) = \sum (-1)^i \operatorname{ch}(E_i) \in A(M)_\mathbb{Q}$ restricts to zero in $A(M \smallsetminus X)_\mathbb{Q}$. Thus $\operatorname{ch}(E_\bullet)$ must be the image of some class in $A_*(X)_\mathbb{Q}$. *MacPherson's graph construction* [4] produces such a localized Chern character $\operatorname{ch}_X^M(E_\bullet)$ in $A_*(X)_\mathbb{Q}$. Then
$$\tau_X(\mathcal{F}) = \operatorname{td}(T_M|_X) \cap \operatorname{ch}_X^M(E_\bullet).$$
The graph construction is an important generalization of the deformation to the normal cone, useful for constructing characteristic classes on their natural loci.

For any scheme X, define the *Todd class* $\operatorname{Td}(X)$ *of* X by
$$\operatorname{Td}(X) = \tau_X(\mathcal{O}_X) \in A_*(X)_\mathbb{Q}.$$
If X is nonsingular, $\operatorname{Td}(X) = \operatorname{td}(T_X) \cap [X]$. Then GRR extends to arbitrary varieties in the following form. If $f \colon X \longrightarrow Y$ is a proper morphism, $\beta \in K^\circ X$, and there is an element $f_*(\beta) \in K^\circ Y$ such that
$$f_*(\beta \otimes [\mathcal{O}_X]) = f_*(\beta) \otimes [\mathcal{O}_Y],$$
then by (1) and (2),
$$f_* \bigl(\operatorname{ch}(\beta) \cap \operatorname{Td}(X) \bigr) = \operatorname{ch}(f_* \beta) \cap \operatorname{Td}(Y).$$

This includes the generalization of Grothendieck's theorem to nonprojective smooth varieties. In addition one has a HRR formula for a vector bundle E on an arbitrary complete variety X,
$$\sum (-1)^i \dim H^i(X, E) = \int_X \operatorname{ch}(E) \cap \operatorname{Td}(X).$$
Another corollary of the general RR theorem is that the induced homomorphism
$$\tau_X \otimes \mathbb{Q}\colon K_\circ(X)_\mathbb{Q} \longrightarrow A_*(X)_\mathbb{Q}$$
is an *isomorphism*, for any algebraic scheme X.[16]

When $f\colon X \longrightarrow Y$ is a regular embedding there are push-forward maps $f_*\colon K^\circ X \longrightarrow K^\circ Y$ and pull-back maps $f^*\colon K_\circ Y \longrightarrow K_\circ X$. Assume that Y can be embedded in a nonsingular variety. Then any locally free sheaf E on X can be resolved on Y, and one sets
$$f_*[E] = \sum (-1)^i [F_i]$$
if $F_\bullet$ is a resolution of $f_* E$, and
$$f^*[\mathcal{F}] = \sum (-1)^i [\mathcal{H}_i(G_\bullet \otimes_Y \mathcal{F})],$$
where $G_\bullet$ is a resolution of $f_* \mathcal{O}_X$, and $\mathcal{H}_i(G_\bullet \otimes_Y \mathcal{F})$ are the homology sheaves of the complex $G_\bullet \otimes_{\mathcal{O}_Y} \mathcal{F}$. Then one has the RR formulas
$$\operatorname{ch} f_*[E] = f_*\bigl(\operatorname{td}(N)^{-1} \cdot \operatorname{ch}(E)\bigr),$$
$$\tau_X f^*[\mathcal{F}] = \operatorname{td}(N)^{-1} \cap f^* \tau_Y(\mathcal{F})$$
for a vector bundle E on X or a coherent sheaf $\mathcal{F}$ on Y; here N is the normal bundle to X in Y. Such formulas were first proved in Grothendieck's seminar SGA 6 and by Verdier [59]. The second gives an *adjunction formula* relating Todd classes:
$$\operatorname{Td}(X) = \operatorname{td}(N)^{-1} \cap f^* \operatorname{Td}(Y).$$

There are similar compatibilities with exterior products. In particular,
$$\operatorname{Td}(X \times Y) = \operatorname{Td}(X) \times \operatorname{Td}(Y).$$

Note that for any complete X,
$$\sum (-1)^i \dim H^i(X, \mathcal{O}_X) = \int_X \operatorname{Td}(X).$$

The above properties of Todd classes generalize classical facts about the arithmetic genus. For example, the constancy of arithmetic genus in flat families generalizes to the fact that Todd classes are compatible with specialization.

For a recent interesting application of the singular Riemann-Roch theorem in local algebra, see Morales [44].[17]

CHAPTER 10

Miscellany

10.1. Topology

For a space X, $H^i X$ denotes the ordinary (singular) cohomology of X, with integer coefficients. For a closed subspace Z of X, $H^i(X, X \smallsetminus Z)$ denotes relative singular cohomology. A useful homology theory for the study of possibly noncompact spaces is the homology with locally finite supports, or Borel-Moore homology, which we denote by $H_i X$. If X is embedded in an oriented real n-manifold M, then

(i) $$H_i X \cong H^{n-i}(M, M \smallsetminus X).$$

Taking $M = \mathbb{R}^n$, this isomorphism may be used to define $H_i X$.[2] If Z is closed in X, and $U = X \smallsetminus Z$, there is a long exact sequence

(ii) $$\cdots \longrightarrow H_{i+1} U \longrightarrow H_i Z \longrightarrow H_i X \longrightarrow H_i U \longrightarrow \cdots$$

and there are cap products

(iii) $$H^i(X, X \smallsetminus Z) \otimes H_j X \xrightarrow{\cap} H_{j-i} Z.$$

Any complex k-dimensional variety V has a *fundamental class* $\operatorname{cl}(V)$ which generates $H_{2k} V \cong \mathbb{Z}$. For a complex variety X there is an induced homomorphism

(iv) $$\operatorname{cl}_X : A_k X \longrightarrow H_{2k} X$$

which takes $\sum n_i [V_i]$ to $\sum n_i \operatorname{cl}(V_i)$. That cl_X respects rational (or algebraic) equivalence is a special case of the proposition which follows.

Any regular embedding $i : X \longrightarrow Y$ of codimension d determines an *orientation class* $u_{X,Y} \in H^{2d}(Y, Y \smallsetminus X)$. If X and Y are nonsingular, $u_{X,Y}$ is determined by the equality $u_{X,Y} \cap \operatorname{cl}(Y) = \operatorname{cl}(X)$. If Y is a vector bundle over X, and i is the zero section, $u_{X,Y}$ is the Thom class of the bundle. For the general case see [4].

PROPOSITION. *Let $i : X \longrightarrow Y$ be a regular embedding of codimension d, V a k-dimensional variety, $f : V \longrightarrow Y$ a morphism and $W = f^{-1}(X)$. Then*

$$f^*(u_{X,Y}) \cap \operatorname{cl}(V) = \operatorname{cl}_W(X \bullet V)$$

in $H_{2k-2d}(W)$.

The proof can be achieved by reducing via familiar methods to the case of divisors, where it is straightforward [**16**, §19]. It follows from this proposition that all our intersection constructions are compatible with those in topology; in particular, the (refined) intersection product on nonsingular varieties maps to the topological product described in §3.1. For varieties meeting properly at a point in a nonsingular variety, it follows that the intersection multiplicity is given by the linking number of the intersections of the varieties with a small sphere about the point.

One may define two cycles α, α' on a complex variety X to be *algebraically equivalent* if there are subvarieties V_i of $X \times C$, C a complete nonsingular curve, and $t_1, t_2 \in C$ with
$$\alpha - \alpha' = \sum [V_i(t_2)] - [V_i(t_1)]$$
(cf. §3.3). If X is nonsingular, one has a filtration of the codimension p cycles on X:
$$\text{Rat}^p X \subset \text{Alg}^p X \subset \text{Hom}^p X \subset \text{Num}^p X \subset Z^p X$$
consisting of cycles rationally, algebraically, homologically and numerically equivalent to zero. For $p = 1$ these groups and the factor groups are quite well understood: $\text{Alg}^1 X / \text{Rat}^1 X$ is the Picard variety, $\text{Alg}^1 X = \text{Hom}^1 X$, $\text{Num}^1 X / \text{Alg}^1 X \cong H^2(X)_{\text{tors}}$ is finite, and $Z^1 X / \text{Num}^1 X$ is finitely generated and free.

For $p > 1$, $Z^p X / \text{Hom}^p X$ is always finitely generated, so $Z^p X / \text{Num}^p X$ is free abelian. We have mentioned that, for surfaces, $\text{Alg}^2 X / \text{Rat}^2 X$ can be "infinite dimensional". Griffiths showed that $\text{Alg}^2 X$ can differ from $\text{Hom}^2 X$ when X is a threefold, and Clemens [11] has improved this to show that $\text{Hom}^2 X / \text{Alg}^2 X$ need not be finitely generated. The principal tool for studying this problem is an Abel-Jacobi map
$$\text{Hom}^p X / \text{Rat}^p X \longrightarrow J^p X$$
to the pth intermediate Jacobian of X.[18]

For $p = 2$, J. P. Murre has recently showed that the image of $\text{Alg}^p X / \text{Rat}^p X$ is an abelian variety, universal for "regular" homomorphisms of $\text{Alg}^p X / \text{Rat}^p X$ to abelian varieties.

10.2. Local complete intersection morphisms

Consider for simplicity the category of varieties that admit closed embeddings into nonsingular varieties. Then any morphism $f \colon X \longrightarrow Y$ admits a factorization $f = pi$
$$X \xrightarrow{i} P \xrightarrow{p} Y$$
with i a closed embedding and p smooth; if $X \subset M$, M nonsingular, one may take $P = Y \times M$, p the projection. We call f a *l.c.i. morphism* if for some (and hence, in fact, for any) such factorization, i is a regular embedding. If p has relative dimension n and i has codimension e, then $d = e - n$ is independent of the factorization, and is the *codimension* of f. In addition, one has the *virtual tangent bundle*
$$T_f = [i^*(T_{P/Y})] - [N_X P] \in K^\circ X.$$
One verifies that such notions are independent of factorization by comparing a factorization through P and through P' with the diagonal factorization:

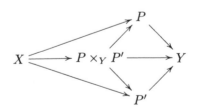

If $f\colon X \longrightarrow Y$ is a l.c.i. morphism of codimension d, f determines *Gysin homomorphisms*
$$f^*\colon A_k X \longrightarrow A_{k-d} X$$
by $f^*\alpha = i^*(p^*\alpha)$, where p^* is flat pull-back (§3.3) and i^* is the Gysin homomorphism for regular embeddings (§5.1). Similarly there are refined Gysin homomorphisms
$$f^!\colon A_k Y' \longrightarrow A_{k-d} X'$$
for any $Y' \longrightarrow Y$, with $X' = X \times_Y Y'$, by $f^*\alpha = i^!(p'^*(\alpha))$, where p' is the induced (flat) morphism from $P \times_Y Y'$ to Y'.

The Riemann-Roch formulas of §9.2 generalize to l.c.i. morphisms $f\colon X \longrightarrow Y$:
$$\operatorname{ch}(f_*\alpha) = f_*\bigl(\operatorname{td}(T_f) \cdot \operatorname{ch}(\alpha)\bigr),$$
$$\tau_X(f^*\beta) = \operatorname{td}(T_f) \cap f^*\tau_Y(\beta)$$
for $\alpha \in K^\circ X$ or $\beta \in K_\circ Y$.

If $i\colon X \longrightarrow Y$ is a regular embedding of codimension d, and $\widetilde{Y}$ is the blow-up of Y along X, then the morphism $f\colon \widetilde{Y} \longrightarrow Y$ is a l.c.i. morphism of codimension zero (by the lemma in §2.4). Consider the fibre square:

$$\begin{array}{ccc} \widetilde{X} & \xrightarrow{j} & \widetilde{Y} \\ {\scriptstyle g}\downarrow & & \downarrow{\scriptstyle f} \\ X & \xrightarrow{i} & Y \end{array}$$

The exceptional divisor $\widetilde{X}$ is $\mathbb{P}(N)$, $N = N_X Y$, and the excess bundle E is the quotient bundle on $\mathbb{P}(N)$;
$$E = g^*N / N_{\widetilde{X}}\widetilde{Y} = g^*N / \mathcal{O}_N(-1).$$

One can show that there are split exact sequences
$$0 \longrightarrow A_k X \xrightarrow{a} A_k\widetilde{X} \oplus A_k Y \xrightarrow{b} A_k\widetilde{Y} \longrightarrow 0$$
with $a(\alpha) = (c_{d-1}(E) \cap g^*\alpha, -i_*\alpha)$ and $b(\beta,\gamma) = j_*\beta + f^*\gamma$. Moreover there is the following general formula for f^*, involving Segre classes. For any k-dimensional subvariety V of Y, let $\widetilde{V}$ be the blow-up of V along $V \cap X$. Then
$$f^*[V] = [\widetilde{V}] + j_*\{c(E) \cap g^*s(V \cap X, V)\}_k$$
in $A_k\widetilde{Y}$. See [**16**, §6.7] for details.

10.3. Contravariant and bivariant theories

We have mentioned the problem of giving a geometric construction of a suitable contravariant ring-valued ("cohomology") theory A^* to pair with the covariant ("homology") theory A_* we have been studying. At present there are several definitions of such rings A^*X, each with its uses as well as defects:

(1) For quasiprojective varieties X, one may define [**4**, Appendix]
$$A^*X = \varinjlim A^*Y,$$

the limit over all morphisms $f\colon X \longrightarrow Y$ from X to nonsingular varieties Y, with A^*Y as in §5.2. There are *pull-backs* $f^*\colon A^*X \longrightarrow A^*X'$ for any morphism $f\colon X' \longrightarrow X$, cap products

$$A^p X \otimes A_q X \longrightarrow A_{q-p} X,$$

with the usual projection formula, and vector bundles have *Chern classes* in A^*X. For complex varieties there are homomorphisms

$$\mathrm{cl}\colon A^*X \longrightarrow H^{2*}X$$

to cohomology. With this theory one also has the desirable properties

$$\mathrm{Pic}(X) \cong A^1 X \quad \text{and} \quad \mathrm{ch}\colon K^\circ X_{\mathbb{Q}} \xrightarrow{\cong} A^* X_{\mathbb{Q}}.$$

However, there are few other functorial properties known. For example, one would like *Gysin homomorphisms*

$$f_*\colon A^p X \longrightarrow A^{p+d} Y$$

for a proper l.c.i. morphism $f\colon X \longrightarrow Y$ of codimension d; it is not clear how to construct such f_* for this theory, even for smooth projections.

Note that if X is nonsingular, A^*X is the same as that constructed in §5.2. It follows that this theory A^* is the *finest* possible contravariant theory agreeing with the given theory on nonsingular varieties.

(2) For any X one may construct an *operational theory* A^*X as follows [**23**], [**16**]: an element c of $A^p X$ is a collection of homomorphisms

$$c_{X'}\colon A_q X' \longrightarrow A_{q-p} X'$$

for all $X' \longrightarrow X$, and all q, compatible with all our other intersection operations. Precisely, one requires that if $f\colon X'' \longrightarrow X'$ is proper (resp. flat, or a regular embedding) with $X' \longrightarrow X$ given, then for $\alpha \in A_*X''$ (resp. $\beta \in A_*X'$)

$$c_{X'}(f_*\alpha) = f_* c_{X''}(\alpha) \qquad (\text{resp. } c_{X''}(f^*\beta) = f^* c_{X'}(\beta)).$$

The ring structure on this A^*X is constructed by composing homomorphisms. This theory has pull-backs, cap products, Chern classes, and also Gysin homomorphisms f_* for l.c.i. morphisms f. However, the map from $\mathrm{Pic}(X)$ to $A^1 X$ need not be an isomorphism, and we do not know a homomorphism from these groups A^*X to cohomology H^*X, for complex varieties X.[19]

These operational groups are useful because of their formal properties. For any series of operations that finally end up with a class in a group A_*X—e.g. for any enumerative problem—there is no loss at all in using them. Even less is known about computations of these A^*X than in the classical case, however. One can at least show that $A^p X = 0$ for $p < 0$ or $p > \dim X$; our proof that A^*X is commutative uses resolution of singularities, so is known only in characteristic zero.

When X is nonsingular, this A^*X also agrees with that in §5.1. Note that this A^* is the *coarsest* theory with this property and with a theory of cap products compatible with intersection products.

(3) Mumford [**46**] has used the image of the first of these theories in the second.

(4) One may define $A^p X$ to be $H^p(X, \mathcal{K}_p)$, where $\mathcal{K}_p$ is Quillen's sheaf of higher K-groups [**52**]. When X is regular, Quillen proved Bloch's formula:

$$H^p(X, \mathcal{K}_p) = A_{n-p} X,$$

$n = \dim(X)$. Gillet [25] has constructed Chern classes in these groups, cap products, and some Gysin homomorphisms.

(5) Another possibility has been proposed by M. Levine, in order to extend results about vector bundles by Murthy and Swan, and Collino, to general singular varieties.

It is most useful if the covariant and contravariant theories are part of a general *bivariant theory* [23]. This should assign to any morphism $f\colon X \longrightarrow Y$ a graded abelian group
$$A^*(X \xrightarrow{f} Y)$$
with *products*, for $f\colon X \longrightarrow Y$, $g\colon Y \longrightarrow Z$,
$$A^p(X \xrightarrow{f} Y) \otimes A^q(Y \xrightarrow{g} Z) \longrightarrow A^{p+q}(X \xrightarrow{gf} Z);$$
push-forwards, for $f\colon X \longrightarrow Y$ proper, $g\colon Y \longrightarrow Z$,
$$f_*\colon A^p(X \xrightarrow{gf} Z) \longrightarrow A^p(Y \xrightarrow{g} Z);$$
and *pull-backs*, for $f\colon X \longrightarrow Y$, $h\colon Y' \longrightarrow Y$,
$$h^*\colon A^p(X \xrightarrow{f} Y) \longrightarrow A^p(X \times_Y Y' \xrightarrow{f'} Y')$$
with $f'\colon X \times_Y Y' \longrightarrow Y'$ the induced morphism. These three options should satisfy various compatibility axioms. Then one sets
$$A^p X = A^p(X \xrightarrow{\mathrm{id}} X), \qquad A_q X = A^{-q}(X \longrightarrow \mathrm{pt.}),$$
where pt. = $\mathrm{Spec}(K)$. The products for the composite $X \xrightarrow{\mathrm{id}} X \xrightarrow{\mathrm{id}} X$ and $X \xrightarrow{\mathrm{id}} X \longrightarrow$ pt. give "cup" and "cap" products.

One point of such a theory is that *orientations* for morphisms $f\colon X \longrightarrow Y$ should determine classes in $A^*(X \xrightarrow{f} Y)$. For example, a flat morphism or a l.c.i. morphism f of codimension d should determine a class $[f]$ in $A^d(X \xrightarrow{f} Y)$. And such a class determines *Gysin homomorphisms*
$$f^*\colon A_k Y \longrightarrow A_{k-d} X, \quad f^*\alpha = [f]\cdot \alpha,$$
$$f_*\colon A^k X \longrightarrow A^{k+d} Y, \quad f_*\alpha = f_*(\alpha \cdot [f]).$$

When a class in A^*X lives naturally on a locus $Z \subset X$—as has been a frequent theme in these lectures—the class should really be a class in $A^*(Z \longrightarrow X)$. For closed embeddings $Z \hookrightarrow X$, $A^*(Z \longrightarrow X)$ should function as local Chow cohomology groups $A^*_Z X$.

In topology there is a natural bivariant theory $H^*(X \xrightarrow{f} Y)$. If one embeds X in $\mathbb{R}^n$, one may define
$$H^i(X \xrightarrow{f} Y) = H^{i+n}(Y \times \mathbb{R}^n, Y \times \mathbb{R}^n \smallsetminus X).$$

At present we have only an *operational* bivariant theory for rational equivalence: a class in $A^p(X \xrightarrow{f} Y)$ is defined to be a collection of homomorphisms from $A_q Y'$ to $A_{q-p}(X \times_Y Y')$ for all $Y' \longrightarrow Y$, all q, compatible as in (2) above. One can show that $A^{-q}(X \longrightarrow \mathrm{pt.})$ is isomorphic to $A_q X$, that there are orientation classes $[f]$ for flat and l.c.i. morphisms, and that our constructions of degeneracy classes, residual intersection classes, local Chern classes, etc., all belong to appropriate bivariant groups.[20]

When $X \longrightarrow Y$ is a closed embedding the local cohomology groups $H_X^p(Y, \mathcal{K}_p)$ used by Gillet look promising for a sharper bivariant theory. D. Grayson has pointed out, however, that for general $f \colon X \longrightarrow Y$, if one embeds X in a nonsingular M, the groups $H_X^*(Y \times M, \mathcal{K}_*)$ are *not* independent of the embedding.

There is a satisfactory bivariant theory specializing to $K_\circ$ and K°, which should agree with the ideal rational equivalence theory $\otimes \mathbb{Q}$. The objects of $K_\circ(X \longrightarrow Y)$ are complexes on X of finite Tor dimension over Y.

On singular varieties the intersection homology theory of Goresky and MacPherson [27] has led to many new insights. One does not know an analogous theory lying between A^*X and A_*X. The place of algebraic cycles in their theory is not very well understood.[21]

10.4. Serre's intersection multiplicity

If two subvarieties V, W of a nonsingular variety X meet properly at a point P, Serre [57] showed that the intersection multiplicity $i(P, V \bullet W; X)$ is given by the formula
$$i(P, V \bullet W; X) = \sum (-1)^i \operatorname{length}\bigl(\operatorname{Tor}_i^A(A/I, A/J)\bigr),$$
where A is the local ring of X at P, and I and J are the ideals of V and W. A unique feature of this formulation is that, at least in its statement, it requires no reduction to the diagonal. This definition makes sense, in fact, for any regular local ring A, whether it contains a field or not. In this generality the *positivity* of this multiplicity remains an open question. Recently, Dutta, Hochster, and MacLaughlin [14] have shown that the natural generalization of this conjecture to modules of finite projective dimension is false, even in the geometric case. In the process they produce some interesting resolutions of modules, which cannot be pull-backs of complexes of vector bundles from any nonsingular variety.

For arbitrary varieties V, W on a nonsingular X, the virtual sheaf
$$\operatorname{Tor}^X(V, W) = \sum (-1)^i [\operatorname{Tor}_i^{\mathcal{O}_X}(\mathcal{O}_V, \mathcal{O}_W)]$$
is an element of $K_\circ(V \cap W)$. One can show that, if τ is the Riemann-Roch homomorphism (§9.2), then
$$\tau(\operatorname{Tor}^X(V, W)) = V \bullet W + \text{lower terms}$$
in $A_*(V \cap W)_\mathbb{Q}$, even in the case of excess intersection.[17]

With Faltings' recent solution of the Mordell conjecture via solutions of conjectures of Tate and Shafarevich, one may anticipate a renewed interest in intersection theory on arithmetic varieties. For such applications it is important to bring in the infinite primes, as in [1].[22]

References

1. S. J. Arakelov, *Theory of intersections on the arithmetic surface*, Proc. Internat. Congress Math. (Vancouver, 1974), vol. 1, Canad. Math. Congr., (Montreal, Quebec), 1975, pp. 405–408.
2. J. K. Arason and A. Pfister, *Quadratische Formen über affinen Algebren und ein algebraischer Beweis des Satzes von Borsuk-Ulam*, J. Reine Angew. Math. **331** (1982), 181–184, Ibid. **339** (1983), 163–164.
3. E. Arbarello, M. Cornalba, P. Griffiths, and J. Harris, *Geometry of algebraic curves*, vol. 1, Springer-Verlag, 1985.
4. P. Baum, W. Fulton, and R. MacPherson, *Riemann-Roch for singular varieties*, Inst. Hautes Études Sci. Publ. Math **45** (1975), 101–167.
5. É. Bézout, *Sur le degré des équations résultantes de l'évanouissement des inconnus*, Mémoires présentés par divers savants à l'Académie des sciences de l'Institut de France (1764).
6. _____, *Théorie générale des équations algébriques*, Ph.D. thesis, Pierres, Paris, 1779.
7. S. Bloch and D. Gieseker, *The positivity of the Chern classes of an ample vector bundle*, Invent. Math. **12** (1971), 112–117.
8. S. Bloch and J. P. Murre, *On the Chow group of certain types of Fano threefolds*, Compositio Math. **39** (1979), 47–105.
9. A. Borel and J.-P. Serre, *Le théorème de Riemann-Roch (d'après Grothendieck)*, Bull. Soc. Math. France **86** (1958), 97–136.
10. C. Chevalley, *Anneaux de Chow et applications*, Secrétariat Math., Paris (1958).
11. H. Clemens, *Homological equivalence, modulo algebraic equivalence, is not finitely generated*, Inst. Hautes Études Sci. Publ. Math. **58** (1984), 19–38.
12. A. Collino, *The rational equivalence ring of symmetric products of curves*, Illinois J. Math. **19** (1975), 567–583.
13. M. Demazure, *Désingularisation des variétés de Schubert généralisées*, Ann. Sci. École. Norm. Sup. **7** (1974), 52–88.
14. S. Dutta, M. Hochster, and J. E. McLaughlin, *Modules of finite projective dimension with negative intersection multiplicities*, Invent. Math. **79** (1985), 253–291.
15. C. Ehresmann, *Sur la topologie de certains espaces homogènes*, Ann. of Math. **35** (1934), 396–443.
16. W. Fulton, *Intersection theory*, Ergeb. Math. Grenzgb., 3 Folge, vol. 2, Springer-Verlag, 1984.
17. W. Fulton and H. Gillet, *Riemann-Roch for general algebraic varieties*, Bull. Soc. Math. France **111** (1983), 287–300.
18. W. Fulton, S. Kleiman, and R. MacPherson, *About the enumeration of contacts*, Springer Lecture Notes in Math. **997** (1983), 156–196.
19. W. Fulton and R. Lazarsfeld, *On the connectedness of degeneracy loci and special divisors*, Acta Math. **146** (1981), 271–283.
20. _____, *Positive polynomials for ample vector bundles*, Ann. of Math. (2) **118** (1983), 35–60.
21. W. Fulton and R. MacPherson, *Intersecting cycles on an algebraic variety*, Real and Complex Singularities, Oslo 1976 (P. Holm, ed.), Sijthoff and Noordhoff, 1977, pp. 179–197.
22. _____, *Defining algebraic intersections*, Algebraic Geometry (Proc. Sympos., Univ. Tromsø, Tromsø, 1977), Springer Lecture Notes in Mathematics **687** (1978), 1–30.
23. _____, *Categorical framework for the study of singular spaces*, Mem. Amer. Math. Soc. **243** (1981).
24. M. Gerstenhaber, *On the deformations of rings and algebras: II*, Ann. of Math. (2) **84** (1966), 1–19.
25. H. Gillet, *Universal cycle classes*, Compositio Math. **49** (1983), 3–49.
26. R. M. Goresky, *Whitney stratified chains and cochains*, Trans. Amer. Math. Soc. **261** (1981), 175–196.
27. M. Goresky and R. MacPherson, *Intersection homology theory*, Topology **19** (1980), 135–162.
28. A. Grothendieck, *La théorie des classes de Chern*, Bull. Soc. Math. France **86** (1958), 137–154.
29. J. Harris and L. Tu, *On symmetric and skew-symmetric determinantal varieties*, Topology **23** (1984), 71–84.

REFERENCES

30. R. Hartshorne, *Algebraic geometry*, Graduate Texts in Math., vol. 52, Springer-Verlag, 1977.
31. H. Hiller, *Geometry of Coxeter groups*, Research Notes in Math., vol. 54, Pitman, New York, 1982.
32. H. Hironaka, *Resolution of singularities of an algebraic variety over a field of characteristic zero*, Ann. of Math. (2) **79** (1964), 109–326.
33. M. Hochster, *Grassmannians and their Schubert subvarieties are arithmetically Cohen-Macaulay*, J. Algebra **25** (1973), 40–57.
34. C. Huneke, *A remark concerning multiplicities*, Proc. Amer. Math. Soc. **85** (1982), 331–332.
35. T. Józefiak, A. Lascoux, and P. Pragacz, *Classes of determinantal varieties associated with symmetric and skew-symmetric matrices*, Math. USSR-Izv. **18** (1982), 575–586.
36. G. Kempf and D. Laksov, *The determinantal formula of Schubert calculus*, Acta Math. **132** (1974), 153–162.
37. S. Kleiman, *The transversality of a general translate*, Compositio Math. **38** (1974), 287–297.
38. E. Kunz, *Einführing in die kommutative algebra und algebraische Geometrie*, Friedr. Vieweg & Sohn, Braunschweig, 1980.
39. A. Lascoux, *Classes de Chern d'une produit tensoriel*, C. R. Acad. Sci. Paris Ser. A **286** (1978), 385–387.
40. R. Lazarsfeld, *Excess intersection of divisors*, Compositio Math. **43** (1981), 281–296.
41. F. S. Macaulay, *Algebraic theory of modular systems*, Cambridge Tracts in Math., Cambridge Univ. Press, 1916.
42. I. G. Macdonald, *Symmetric functions and Hall polynomials*, Oxford Univ. Press, 1979.
43. E. Martinelli, *Sulla varietà delle faccette p-dimensionali di S_r*, Atti. Accad. Italia Mem. Cl. Sci. Fis. Mat. Nat. **12** (1941), 917–943.
44. M. Morales, *Polynôme d'Hilbert-Samuel des clôtures intégrales des puissances d'un ideal m-primaire*, Bull. Sci. Math. France **112** (1984), 343–358.
45. D. Mumford, *Rational equivalences of 0-cycles on surfaces*, J. Math. Kyoto Univ. **9** (1969), 195–204.
46. _____, *Towards an enumerative geometry of the moduli space of curves*, Arithmetic and Geometry: Papers dedicated to I. R. Shafarevich (M. Artin and J. Tate, eds.), vol. II, Birkhäuser, 1983, pp. 271–328.
47. V. Navarro Aznar, *Sur les multiplicités de Schubert locales des faisceaux algébriques cohérents*, Compositio Math. **48** (1983), 311–326.
48. I. Newton, *Geometrica analytica*, 1680.
49. M. Pieri, *Formule di coincidenza per le serie algebriche ∞^n di coppie di punti dello spazio a n dimensioni*, Rend. Circ. Mat. Palermo **5** (1891), 252–268.
50. J. Plücker, *Solution d'une question fondamentale concernant la théorie générale des courbes*, J. Reine Angew. Math. **12** (1834), 105–108.
51. J. V. Poncelet, *Traité des propriétés projectives des figures, 1822*, Gauthier-Villars, Paris, 1865.
52. D. Quillen, *Higher algebraic K-theory: I*, Springer Lecture Notes in Math. **341** (1973), 85–147.
53. G. Salmon, *On the degree of a surface reciprocal to a given one*, Cambridge and Dublin Math. J. **2** (1847), 65–73.
54. P. Samuel, *La notion de multiplicité en algèbre et en géométrie algébrique*, J. Math. Pures Appl. **30** (1951), 159–274.
55. B. Segre, *On limits of algebraic varieties, in particular of their intersections and tangential forms*, Proc. London Math. Soc. **47** (1942), 351–403.
56. _____, *Nuovi metodi e resultati nella geometria sulle varietà algebriche*, Ann. Mat. Pura Appl. (4) **35** (1953), 1–128.
57. J.-P. Serre, *Algèbre locale. Multiplicités: Cours au Collège de France, 1957/58*, 2nd ed., Springer Lecture Notes in Math. **11** (1965).
58. F. Severi, *Il concetto generale di multiplicità delle soluzioni pei sistemi di equazioni algebriche e la teoria dell'eliminazione*, Ann. Mat. Pura Appl. (4) **26** (1947), 221–270.
59. J.-L. Verdier, *Le théorème de Riemann-Roch pour les intersections complètes*, Astérisque **36–37** (1976), 189–228.
60. B. L. van der Waerden, *Modern algebra, I, II*, Ungar, New York, 1950.
61. A. Weil, *Lectures*, Institute for Advanced Study, 1981–1982.
62. A. Zobel, *On the non-specialization of intersection on a singular variety*, Mathematika **8** (1961), 39–44.

Notes (1983–1995)

1. (p. 5) For a description of the intersection ring of the space of complete quadrics, see

   C. De Concini and C. Procesi, *Complete symmetric varieties, II. Intersection theory*, in *Algebraic Groups and Related Topics*, Advanced Studies in Pure Math., vol. 6, North-Holland, 1985, pp. 481–513.

   The Chow ring of these varieties is still only partially understood.

2. (pp. 20, 69) An elementary construction of this fundamental class, following [**23**], is given in Appendix B of

   W. Fulton, *Young tableaux, with applications to representation theory and geometry*, Cambridge University Press, to appear.

3. (p. 23) Grothendieck had proved this in [**10**] under the weaker assumption that E is an affine bundle over X. Gillet proved it with no group acting on the bundle, in

   H. Gillet, *Riemann-Roch theorems for higher algebraic K-theory*, Advances in Math. **40** (1981), 203–289.

   For an application to a stronger splitting principle for Chow groups, see the second reference in Note 10.

4. (p. 33) For more along these lines, see

   R. Smith and R. Varley, *Singularity theory applied to Θ-divisors*, Springer Lecture Notes in Mathematics **1479** (1991), 238–257.

5. (p. 39) Another proof of this functoriality can be found in

   A. Vistoli, *Intersection theory on algebraic stacks and on their moduli spaces*, Invent. Math. **97** (1989), 613–670.

6. (p. 41) Although computing Chow groups and rings of general smooth projective varieties remains a very hard problem, there are now many more varieties about which something is known. A careful survey of this could take a volume by itself. Here is a small sampling of references:

A. Beauville, *Sur l'anneau de Chow d'une variété abélienne*, Math. Annalen **273** (1986), 647–651.

G. Ellingsrud and S. A. Strømme, *On the Chow ring of a geometric quotient*, Ann. of Math. **130** (1989), 159–187.

A. Collino and W. Fulton, *Intersection rings of spaces of triangles*, Mém. Bull. Soc. Math. France **117** (1989), 75–117.

C. Faber, *Chow rings of moduli spaces of curves. I. The Chow ring of $\overline{M}_3$; II. Some results on the Chow ring of $\overline{M}_4$*, Ann. of Math. **132** (1990), 331–419, 421–449.

G. Ellingsrud and S. A. Strømme, *Towards the Chow ring of the Hilbert scheme of $\mathbb{P}^2$*, J. Reine Angew. Math. **441** (1993), 33–44.

S. Keel, *Intersection theory of moduli space of stable n-pointed curves of genus zero*, Trans. Amer. Math. Soc. **330** (1992), 545–574.

K. H. Paranjape, *Cohomological and cycle-theoretic connectivity*, Ann. of Math. **139** (1994), 641–660.

W. Fulton, R. MacPherson, F. Sottile, and B. Sturmfels, *Intersection theory on spherical varieties*, J. Alg. Geom. **4** (1995), 181–193.

Many other calculations of Chow groups are contained in other papers mentioned elsewhere in these notes.

7. (p. 44) For higher degrees, it is still the case that only a few of these numbers are known. For some modern work on this, see

> P. Aluffi, *The enumerative geometry of plane cubics I: smooth cubics*, Trans. Amer. Math. Soc. **317** (1990), 501–539.
>
> S. Kleiman and R. Speiser, *Enumerative geometry of nonsingular plane cubics*, in *Algebraic Geometry: Sundance 1988*, Contemp. Math. **116** (1991), 85–113.

8. (p. 50) These formulas are now special cases of a general formula for degeneracy loci of maps between two bundles with flags of subbundles. There is such a locus for each permutation, and the corresponding formula is given by the corresponding "double Schubert polynomial" of Lascoux and Schützenberger. The proof of the general formula is easier than those described here, in that it requires only a knowledge of $\mathbb{P}^1$-bundles in place of the calculations of Gysin formulas. For details, see

> W. Fulton, *Flags, Schubert polynomials, degeneracy loci, and determinantal formulas*, Duke Math. J. **65** (1992), 381–420.

9. (p. 51) For this, see

> P. Pragacz, *Cycles of isotropic subspaces and formulas for symmetric degeneracy loci*, in *Topics in Algebra*, Banach Center Publications, vol. 26, part 2, 1990, pp. 189–199.

10. (p. 52) As in Note 8, these formulas have become part of a more general story of degeneracy loci. For each of the classical groups, there is such a locus for each element in the corresponding Weyl group. For this, see

> W. Fulton, *Determinantal formulas for orthogonal and symplectic degeneracy loci*, to appear in J. Diff. Geom.

> W. Fulton, *Schubert varieties in flag bundles for the classical groups*, to appear in *Proceedings of Conference in Honor of Hirzebruch's 65th Birthday*, Bar Ilan, 1993, Amer. Math. Soc.

> P. Pragacz and J. Ratajski, *Formulas for Lagrangian and orthogonal degeneracy loci; the $\widetilde{Q}$-polynomials approach*, preprint.

The second reference includes the deduction of the general case from the case when L is a square; as Totaro points out, this deduction is not as simple as had been thought, since there is no "squaring principle" for line bundles that includes 2-torsion.

11. (p. 54) In case one is intersecting with divisors in one linear system, it is possible to find a further refinement of these intersection products, at a possible cost of extending the ground field. For the strongest results in this direction, see

> L. van Gastel, *Excess intersections and a correspondence principle*, Invent. Math. **103** (1991), 197–211.

Vogel and his coauthors have continued to study the refinements of Bézout's theorem. For example, see

> H. Flenner and W. Vogel, *Improper intersections and a converse to Bezout's theorem*, J. of Algebra **159** (1993), 460–476.

In case the ambient variety is projective space, the paper of van Gastel includes an explanation of how to translate between the constructions of Vogel and the intersection theory described in this book.

12. (p. 55) In fact, all of the conics can be real! We discovered this in 1986, but did not publish a proof. Recently a detailed proof has been given:

> F. Ronga, A. Tognoli, and T. Vust, *The number of conics tangent to five given conics: the real case*, preprint.

F. Sottile, in his 1994 University of Chicago PhD thesis, proved analogous results for intersections of Schubert cycles in any Grassmannian of lines in any projective space. The methods in all cases are by explicit deformations. It is intriguing to speculate about how general this phenomenon is, when the problem is one of counting how many figures of some kind have a given position with respect to some given general figures.

13. (p. 56) The general case of this has now been proved:

S. Kleiman, J. Lipman, and B. Ulrich, *The source double-point cycle of a finite map of codimension one*, in *Complex Projective Geometry*, London Math. Soc. Lecture Note Series **179** (1992), 199–212.

14. (p. 57)   For more about multiple point formulas, see

    S. Kleiman, *Multiple point formulas I: iteration*, Acta Math. **147** (1981), 13–49.

    S. Kleiman, *Multiple point formulas II: the Hilbert scheme*, in *Enumerative Geometry (Sitges, 1987)*, Springer Lecture Notes in Math. **1436** (1990), 101–138.

15. (p. 61)   For a generalization, see

    W. Fulton, *Positive polynomials for filtered ample vector bundles*, Amer. J. Math. **117** (1995), 627–633.

16. (p. 68)   For a deduction of the singular case from the nonsingular case, see

    B. Angeniol and F. El Zein, *Théorème de Riemann-Roch par désingularisation*, Bull. Sci. Math. France **116** (1988), 385–400.

17. (pp. 68, 74)   P. Roberts has used these ideas, especially the graph construction, to prove part of a conjecture of Serre about the vanishing of the intersection number in local algebra:

    P. Roberts, *Local Chern characters and intersection multiplicities*, in *Algebraic Geometry, Bowdoin 1985*, Proc. Sympos. Pure Math. **46** part 2, Amer. Math. Soc., 1987, pp. 389–400.

An independent proof was also given by Gillet and Soulé using K-theory and Adams operations:

    H. Gillet and C. Soulé, *Intersection theory using Adams operations*, Invent. Math. **90** (1987), 243–277.

More on the graph construction can be found in:

    H. Gillet and C. Soulé, *An arithmetic Riemann-Roch theorem*, Invent. Math. **110** (1992), 493–543.

18. (p. 70)   There has been considerable progress on the relations between cycles and intermediate Jacobians. For example:

    C. Voisin, *Une approche infinitesimal du théorème de H. Clemens sur les cycles d'une quintique génerale de $\mathbb{P}^4$*, J. Algebraic Geometry **1** (1992), 157–174.

    M. Nori, *Algebraic cycles and Hodge-theoretic connectivity*, Invent. Math. **111** (1993), 349–373.

N. Suwa, *Sur l'image de l'application d'Abel-Jacobi de Bloch*, Bull. Sci. Math. France **116** (1988), 69–101.

H. Esnault and M. Levine, *Surjectivity of cycle maps*, in *Journées de Géométrie Algébrique d'Orsay*, Asterisque **218** (1993), 203–216.

19. (p. 72) Totaro has shown by examples why this cannot exist in general:

B. Totaro, *Chow groups, Chow cohomology, and linear varieties*, to appear in J. Alg. Geom.

20. (p. 73) Kleiman and Thorop have given some variations on this theme, in section 3 of

S. Kleiman, *Intersection theory and enumerative geometry; a decade in review*, Proc. Symp. Pure Math. Amer Math Soc. **46** (2), 1987, pp. 321–370.

Practical methods for calculating these groups have also been given:

S. Kimura, *Fractional intersection and bivariant theory*, Communications in Algebra **20** (1992), 285–302.

Kimura's paper also explains how rational intersection numbers for curves on normal surfaces can be interpreted by means of these operational Chow cohomology groups. For another approach to Chow cohomology, see

A. Suslin and V. Voevodsky, *Relative cycles and Chow sheaves*, preprint.

21. (p. 74) There has been some progress on this question:

G. Barthel, J.-P. Brasselet, K.-H. Fieseler, O. Gabber, and L. Kaup, *Relèvement de cycles algébriques et homomorphismes associés en homologie d'intersection*, Ann. of Math. **141** (1995), 147–179.

22. (p. 74) This development has taken place and is continuing. A general theory has been developed:

H. Gillet and C. Soulé, *Arithmetic intersection theory*, Inst. Hautes Études Sci. Publ. Math. **72** (1991), 94–174.

For a survey, with references, see

C. Soulé, D. Abramovich, J.-F. Burnol, and J. Kramer, *Lectures on Arakelov Geometry*, Cambridge Studies in Advanced Mathematics, vol. 33, Cambridge University Press, 1992.

These notes have only described some recent work in intersection theory as it relates to topics discussed in the book. There have also been several important developments that go well beyond what was envisioned in 1984. We mention only a few of these, with a small sampling of references:

Bloch's higher Chow groups, and relations to higher K-theory and Beilinson's regulator:

>S. Bloch, *Algebraic cycles and higher K-theory*, Adv. in Math. 61 (1986), 267–304.

>C. Deninger, *The Beilinson conjectures*, in *L-functions and Arithmetic (Durham, 1989)*, London Math. Soc. Lecture Note Series **153** (1991), 173–209.

>H. Esnault and E. Viehweg, *Deligne-Beilinson cohomology*, in *Beilinson's Conjectures on Special Values of L-functions*, Perspectives in Mathematics, vol. 4, Academic Press, 1988, pp. 43–92.

Motives and Chow groups:

>J.-P. Murre, *On the motive of an algebraic surface*, J. Reine Angew. Math. **409** (1990), 190–204.

>U. Jannsen, *Motivic sheaves and filtrations on Chow groups*, in *Motives (Seattle WA 1991)*, Proc. Sympos. Pure Math. **55**, part 1, 1994, pp. 245–302.

>V. Voevodsky, *Triangulated categories of motives over a field*, preprint.

A theory, based on homotopy groups of Chow varieties, called Lawson homology, that interpolates between ordinary homology and Chow groups of algebraic varieties:

>H. B. Lawson, *Algebraic cycles and homotopy theory*, Ann. of Math. **129** (1989), 253–291.

>E. M. Friedlander and H. B. Lawson, *A theory of algebraic cocycles*, Ann. of Math. 136 (1992), 361–428.

Intersection theory on moduli spaces, especially as influenced by physics:

>E. Witten, *Two-dimensional gravity and intersection theory on moduli space*, in *Surveys in Differential Geometry (Cambridge, MA, 1990)*, Lehigh Univ., 1991, pp. 243–310.

>M. Kontsevich, *Intersection theory on the moduli space of curves*, Funct. Anal. and Appl. **25** (1991), 123–129.

>E. Looijenga, *Intersection theory on Deligne-Mumford compactifications (after Witten and Kontsevich)*, Séminaire Bourbaki, Exp. 768, 1992–93, Asterisque **216** (1993), 187–212.

Quantum cohomology, with applications to enumerative geometry:

>M. Kontsevich and Yu. Manin, *Gromov-Witten classes, quantum cohomology, and enumerative geometry*, Comm. Math. Phys. **164** (1994), 525–562.

>P. Di Francesco and C. Itzykson, *Quantum intersection rings*, preprint.

W. Fulton and R. Pandharipande, *Notes on stable maps and quantum cohomology*, preprint.

Kleiman and Thorup (see Note 20) defined the notion of an Alexander scheme. The Chow groups of such a variety, at least after tensoring with $\mathbb{Q}$, have a natural ring structure. These are now quite well understood:

A. Vistoli, *Alexander duality in intersection theory*, Compositio Math. **70** (1989), 199–225.

S. Kimura, *On the characterization of Alexander schemes*, Compositio Math. **92** (1994), 273–284.

Most of the intersection theory described here has been extended to Deligne-Mumford stacks: See Vistoli (Note 5) and

H. Gillet, *Intersection theory on algebraic stacks and Q-varieties*, J. Pure and Appl. Algebra **39** (1984), 193–240.

Finally—with apologies to the many whose papers should be included in such a list—a few other papers that may be of interest to readers of these notes:

S. Bloch, M. P. Murthy, and L. Szpiro, *Zero-cycles and the number of generators of an ideal*, Mém. Soc. Math. France **38** (1989), 51–74.

J.-P. Demailly, *Monge-Ampère operators, Lelong numbers and intersection theory*, Complex Analysis and Geometry, Plenum, 1993, pp. 115–193.

S. Keel, *Intersection theory of linear embeddings*, Trans. Amer. Math. Soc. **335** (1993), 195–212.

X. Wu, *Residual intersections and some applications*, Duke Math. J. **75** (1994), 733–758.

P. Aluffi, *Singular schemes of hypersurfaces*, preprint.

There have been many interesting and important papers on enumerative goemetry besides those mentioned in these notes. Some of these can be found in Kleiman's survey in Note 20, but it would take another volume to describe the work in this area during the succeeding decade.

Other Titles in This Series

(Continued from the front of this publication)

53 **Wilhelm Klingenberg,** Closed geodesics on Riemannian manifolds, 1983
52 **Tsit-Yuen Lam,** Orderings, valuations and quadratic forms, 1983
51 **Masamichi Takesaki,** Structure of factors and automorphism groups, 1983
50 **James Eells and Luc Lemaire,** Selected topics in harmonic maps, 1983
49 **John M. Franks,** Homology and dynamical systems, 1982
48 **W. Stephen Wilson,** Brown-Peterson homology: an introduction and sampler, 1982
47 **Jack K. Hale,** Topics in dynamic bifurcation theory, 1981
46 **Edward G. Effros,** Dimensions and C^*-algebras, 1981
45 **Ronald L. Graham,** Rudiments of Ramsey theory, 1981
44 **Phillip A. Griffiths,** An introduction to the theory of special divisors on algebraic curves, 1980
43 **William Jaco,** Lectures on three-manifold topology, 1980
42 **Jean Dieudonné,** Special functions and linear representations of Lie groups, 1980
41 **D. J. Newman,** Approximation with rational functions, 1979
40 **Jean Mawhin,** Topological degree methods in nonlinear boundary value problems, 1979
39 **George Lusztig,** Representations of finite Chevalley groups, 1978
38 **Charles Conley,** Isolated invariant sets and the Morse index, 1978
37 **Masayoshi Nagata,** Polynomial rings and affine spaces, 1978
36 **Carl M. Pearcy,** Some recent developments in operator theory, 1978
35 **R. Bowen,** On Axiom A diffeomorphisms, 1978
34 **L. Auslander,** Lecture notes on nil-theta functions, 1977
33 **G. Glauberman,** Factorizations in local subgroups of finite groups, 1977
32 **W. M. Schmidt,** Small fractional parts of polynomials, 1977
31 **R. R. Coifman and G. Weiss,** Transference methods in analysis, 1977
30 **A. Pełczyński,** Banach spaces of analytic functions and absolutely summing operators, 1977
29 **A. Weinstein,** Lectures on symplectic manifolds, 1977
28 **T. A. Chapman,** Lectures on Hilbert cube manifolds, 1976
27 **H. Blaine Lawson, Jr.,** The quantitative theory of foliations, 1977
26 **I. Reiner,** Class groups and Picard groups of group rings and orders, 1976
25 **K. W. Gruenberg,** Relation modules of finite groups, 1976
24 **M. Hochster,** Topics in the homological theory of modules over commutative rings, 1975
23 **M. E. Rudin,** Lectures on set theoretic topology, 1975
22 **O. T. O'Meara,** Lectures on linear groups, 1974
21 **W. Stoll,** Holomorphic functions of finite order in several complex variables, 1974
20 **H. Bass,** Introduction to some methods of algebraic K-theory, 1974
19 **B. Sz.-Nagy,** Unitary dilations of Hilbert space operators and related topics, 1974
18 **A. Friedman,** Differential games, 1974
17 **L. Nirenberg,** Lectures on linear partial differential equations, 1973
16 **J. L. Taylor,** Measure algebras, 1973
15 **R. G. Douglas,** Banach algebra techniques in the theory of Toeplitz operators, 1973
14 **S. Helgason,** Analysis on Lie groups and homogeneous spaces, 1972
13 **M. Rabin,** Automata on infinite objects and Church's problem, 1972

(See the AMS catalog for earlier titles)